MAXIMUM LIKELIHOOD ESTIMATION WITH STATA

William Gould & William Sribney

College Station, Texas

Stata Press, 702 University Drive East, College Station, Texas 77840

Typeset in TEX
Printed in the United States of America
10 9 8 7 6 5 4 3 2 1

ISBN 1-881228-41-X

Contents

APPENDICES

Preface

Maximum Likelihood Estimation with Stata is written for researchers in all disciplines who need to compute maximum likelihood estimators that are not available as packaged routines. The reader is assumed to be familiar with Stata, but no special programming skills are assumed, except in the last chapter, which details how one might add a new estimation command to Stata. No special theoretical knowledge is needed either, other than an understanding of the likelihood function that will be maximized.

Stata's `ml` command has been completely rewritten in Stata 6, making it easier to set up a problem and code a likelihood function. At its simplest, implementing a maximum likelihood estimator using `ml` consists of three steps:

1. Write a Stata program to compute the log likelihood function. Helper subroutines are included to make this task as painless as possible and to make the resulting program as bulletproof as possible.
2. Use the `ml model` command to define the particular model to estimate. The complete specifications for the problem are set up in a single Stata command.
3. Use the `ml maximize` command to search for good starting values and then carry out the maximization.

More sophisticated variations on this theme are possible: The user-written program that computes the log-likelihood function can also provide analytic first derivatives, and, perhaps, analytic second derivatives as well. Subroutines in `ml` will check that the analytic derivatives are correct by comparing them with numerical derivatives.

Initial values can be provided by the user, or an extensive random/deterministic search for good initial values can be carried out by `ml` with or without guidance from the user.

With little or no additional programming effort, the robust variance estimator for appropriate likelihood functions can be computed by `ml` in place of the traditional variance estimates given by the inverse of the negative second derivatives.

The new `ml` command also does a better job with difficult likelihood functions—functions that are not globally concave. The default optimizer now converges more rapidly and more robustly with such functions. If it runs into problems, the user can redo (or restart) the maximization and specify the `difficult` option, which steps through the nonconcave regions of the likelihood function using a more sophisticated (albeit more time-consuming) algorithm.

The first chapter of this book provides a general overview of maximum likelihood estimation theory and numerical optimization methods with an emphasis on practical implications of each for applied work.

The second chapter is an overview of the `ml` command. Chapters 3–11 detail, step by step, the use of Stata to maximize user-written likelihood functions. Included in the examples are probit, linear regression, Weibull regression, random-effects linear regression for panel data, and the Cox proportional hazards model. Chapter 12 describes how to package all the user-written code in a do-file, so it can be conveniently reused to estimate different models. Chapter 13, the last chapter, details how to structure the code in an ado-file to create a new Stata estimation command.

Appendices contain full syntax diagrams for all the `ml` subroutines and useful checklists for implementing each of the different maximization methods.

William Gould
William Sribney
College Station, Texas
December 1998

1 Theory and Practice

Contents

1.1 Introduction

Stata can estimate user-defined models using the maximum likelihood method and Stata's `ml` command does that. `ml` has a formidable syntax diagram (see Appendix A) but is surprisingly easy to use. Here we use it to implement probit models and estimate a particular one:

```
program define myprobit
        version 6
        args lnf theta
        quietly replace `lnf' = ln(normprob(`theta')) if $ML_y1==1
        quietly replace `lnf' = ln(normprob(-`theta')) if $ML_y1==0
end
```

```
. ml model lf myprobit (foreign=mpg weight)
. ml maximize
initial:       log likelihood = -51.292891
alternative:   log likelihood = -45.055272
rescale:       log likelihood = -45.055272
Iteration 0:   log likelihood = -45.055272
Iteration 1:   log likelihood = -27.904114
Iteration 2:   log likelihood =  -26.85781
Iteration 3:   log likelihood = -26.844198
Iteration 4:   log likelihood = -26.844189
Iteration 5:   log likelihood = -26.844189
                                                   Number of obs   =         74
                                                   Wald chi2(2)    =      20.75
Log likelihood = -26.844189                        Prob > chi2     =     0.0000
------------------------------------------------------------------------------
 foreign |      Coef.   Std. Err.       z     P>|z|       [95% Conf. Interval]
---------+--------------------------------------------------------------------
     mpg |  -.1039503   .0515689     -2.016   0.044      -.2050235   -.0028772
  weight |  -.0023355   .0005661     -4.126   0.000       -.003445   -.0012261
   _cons |   8.275464   2.554142      3.240   0.001       3.269438    13.28149
------------------------------------------------------------------------------
```

We entered a four-line program and then typed two more lines. That was all that was required to program and to estimate a probit model. Moreover, our four-line program is good enough to estimate any probit model. Reported are conventional, inverse negative second derivative variance estimates but, by specifying an option, we could obtain the Huber/White/sandwich robust variance estimates without changing our program.

We will discuss `ml` and how to use it soon—in Chapter 2—but we will discuss the theory and practice of maximizing likelihood functions first.

We will discuss theory so we can use terms like "conventional, inverse negative second derivative variance estimates" and "Huber/White/sandwich robust variance estimates" and you will understand not only what the terms mean but some of the theory underlying them.

We will discuss practice because of what a frustrating experience it can be. A little understanding of how numerical optimizers work goes a long way toward reducing that frustration. A knowledgeable person can glance at output and conclude that better starting values are needed, or that more iterations are needed, or that even though the software reported convergence, the process has not converged.

We will discuss both these issues so that, when things go well, you will know what you have and, when things go badly, you will have some idea of what to do.

1.2 The likelihood maximization problem

Maximum likelihood estimates of the parameter vector $\mathbf{b}$ are found by obtaining

$$\max_{\mathbf{b}} \quad L(\mathbf{b};\mathbf{X})$$

where $\mathbf{X}$ is the data. Most texts will note that this is equivalent to

$$\max_{\mathbf{b}} \quad \ln L(\mathbf{b};\mathbf{X})$$

because maximizing $\ln L()$ is equivalent to maximizing $L()$ because $L()$ is a positive function and $\ln()$ is a monotone increasing transformation. In this notation, $L()$ is the likelihood of the data. Textbooks typically introduce the assumption that "observations" are independent and identically distributed (*i.i.d.*) and rewrite the likelihood as

$$\max_{\mathbf{b}} \quad L(\mathbf{b};\mathbf{X}) = \max_{\mathbf{b}} \quad \ell(\mathbf{b};\mathbf{x}_1) \times \ell(\mathbf{b};\mathbf{x}_2) \times \cdots \times \ell(\mathbf{b};\mathbf{x}_N)$$

or equivalently

$$\max_{\mathbf{b}} \quad \ln L(\mathbf{b};\mathbf{X}) = \max_{\mathbf{b}} \quad \ln \ell(\mathbf{b};\mathbf{x}_1) + \ln \ell(\mathbf{b};\mathbf{x}_2) + \cdots + \ln \ell(\mathbf{b};\mathbf{x}_N)$$

Why do we take logarithms?

1. Speaking statistically, we know how to take expectations (and variances) of sums and it is easy when the individual terms are independent.
2. Speaking numerically, we can make calculations that would be impossible were we not to take logs. That is, we would want to take logs even if logs were not, in the first sense, natural.

To better understand the second point, let us consider a discrete likelihood function, meaning that the likelihoods correspond to probabilities—logit and probit models would be examples. In such cases,

$$\ell(\mathbf{b};\mathbf{x}_j) = \Pr(\text{we would observe } \mathbf{x}_j)$$

where $\mathbf{x}_j$ includes both dependent and independent variables. For instance, $\ell(\mathbf{b};\mathbf{x}_j)$ might be the probability that $x_{j1} = 1$ conditional on x_{j2}, x_{j3}, ..., x_{jk}. The overall likelihood function is then the probability that we would observe the entire set of data and

$$L(\mathbf{b};\mathbf{X}) = \ell(\mathbf{b};\mathbf{x}_1) \times \ell(\mathbf{b};\mathbf{x}_2) \times \cdots \times \ell(\mathbf{b};\mathbf{x}_N)$$

because the N observations are assumed to be independent. Said differently,

$$\text{Pr(dataset)} = \text{Pr(datum 1)} \times \text{Pr(datum 2)} \times \cdots \times \text{Pr(datum } N)$$

Probabilities are bounded by 0 and 1. In the simple probit or logit case, we can hope that $\ell(\mathbf{b};\mathbf{x}_j) > .5$ for almost all j, but that may not be true. If there were many possible outcomes—e.g., multinomial logit—it would be unlikely that $\ell(\mathbf{b};\mathbf{x}_j)$ would be greater than .5. Anyway, let's pretend we are lucky and $\ell(\mathbf{b};\mathbf{x}_j)$ is right around .5 for all N observations. What would be the value of $L()$ if we had, say, 500 observations? It would be

$$.5^{500} \approx 3 \times 10^{-151}$$

That is a very small number. What if we had 1,000 observations? The likelihood would be

$$.5^{1000} \approx 9 \times 10^{-302}$$

What if we had 2,000 observations? The likelihood would be

$$.5^{2000} \approx \texttt{<COMPUTER UNDERFLOW>}$$

Mathematically we can calculate it; it would be roughly 2×10^{-603}, but that number is too small for most digital computers. Modern computers can process a range of roughly 10^{-301} to 10^{301}.

To wit: If we were considering maximum likelihood probit or logit estimators and if we implemented our likelihood function in natural units, we would be unable to deal with more than about 1,000 observations! Taking logs is how programmers solve such problems because logs remap small positive numbers to the entire range of negative numbers. In logs,

$$\ln(.5^{1000}) = 1000 \times \ln(.5) \approx 1000 \times -.6931 = -693.1$$

a number well within computational range. Similarly, 2,000 observations is not a problem:

$$\ln(.5^{2000}) = 2000 \times \ln(.5) \approx 2000 \times -.6931 = -1386.2$$

1.3 Likelihood theory

Let us pretend that, through some miracle, we obtain $\mathbf{b}$ satisfying

$$\max_{\mathbf{b}} \quad L(\mathbf{b};\mathbf{X})$$

or, if you prefer,

$$\max_{\mathbf{b}} \quad \ln \ell(\mathbf{b};\mathbf{x}_1) + \ln \ell(\mathbf{b};\mathbf{x}_2) + \cdots + \ln \ell(\mathbf{b};\mathbf{x}_N)$$

The estimated variation of $\mathbf{b}$ is given by $-\mathbf{H}^{-1}$, where $\mathbf{H}$ is the Hessian (matrix of second derivatives):

$$\mathbf{H} = \frac{\partial^2 \ln L(\mathbf{b};\mathbf{X})}{\partial\mathbf{b}\partial\mathbf{b}'}$$
$$= \frac{\partial^2 \ln \ell(\mathbf{b};\mathbf{x}_1)}{\partial\mathbf{b}\partial\mathbf{b}'} + \cdots + \frac{\partial^2 \ln \ell(\mathbf{b};\mathbf{x}_N)}{\partial\mathbf{b}\partial\mathbf{b}'}$$

Thus, the square root of the diagonal of $-\mathbf{H}^{-1}$ are the estimated standard errors. Do you know why? To show this, let us simplify the notation. Let

$$D = \frac{\partial}{\partial\mathbf{b}} \qquad \text{and} \qquad D^2 = \frac{\partial^2}{\partial\mathbf{b}\partial\mathbf{b}'}$$

so that using this notation, we can rewrite the definition of $\mathbf{H}$ more succinctly as

$$\mathbf{H} = D^2 \ln L(\mathbf{b};\mathbf{X}) = D^2 \ln \ell(\mathbf{b};\mathbf{x}_1) + \cdots + D^2 \ln \ell(\mathbf{b};\mathbf{x}_N)$$

Let us also agree to simply write L and ℓ_j rather than $L(\mathbf{b};\mathbf{X})$ and $\ell(\mathbf{b};\mathbf{x}_j)$ when the meaning would be clear. We will do that with other functions we define along the way, too.

It can be proven that

$$\textbf{Lemma:} \qquad E(D \ln L) = \mathbf{0} \qquad \text{and} \qquad -E(D^2 \ln L) = E\big((D \ln L)^2\big) \tag{1}$$

where $E()$ denotes expectation. One can prove the above Lemma if one is willing to assume $L()$ is the true density function of $\mathbf{X}$. We provide the proof in a technical note that follows shortly, but for now, let's just accept the Lemma as true.

$D \ln L$ is so important that it is given a name—the score vector or gradient vector—and its own notation:

$$\mathbf{g}(\mathbf{b};\mathbf{X}) = D \ln L(\mathbf{b};\mathbf{X}) = \frac{\partial \ln L(\mathbf{b};\mathbf{X})}{\partial\mathbf{b}}$$

Thus, our Lemma states

$$\textbf{Same Lemma:} \qquad E(\mathbf{g}) = \mathbf{0} \qquad \text{and} \qquad -E(\mathbf{H}) = E\big(\mathbf{g}\mathbf{g}'\big)$$

where $\mathbf{g}\mathbf{g}'$ is just the vector equivalent of squaring. The variance of $\mathbf{g}$ is

$$\begin{aligned} \text{Var}(\mathbf{g}) &= E(\mathbf{g}\mathbf{g}') - E(\mathbf{g})\,E(\mathbf{g})' && \text{by definition} \\ &= E(\mathbf{g}\mathbf{g}') && \text{because } E(\mathbf{g}) = \mathbf{0} \text{ (Lemma part 1)} \\ &= -E(\mathbf{H}) && \text{because } -E(\mathbf{H}) = E\big(\mathbf{g}\mathbf{g}'\big) \text{ (Lemma part 2)} \end{aligned} \tag{2}$$

There is another way we could obtain $\text{Var}(\mathbf{g})$: $\mathbf{g}$ is a function of $\mathbf{b}$, so by the delta method we know that

$$\text{Var}(\mathbf{g}) \approx (D\mathbf{g})\,\text{Var}(\mathbf{b})\,(D\mathbf{g})' = (D^2 \ln L)\,\text{Var}(\mathbf{b})\,(D^2 \ln L)' = \mathbf{H}\,\text{Var}(\mathbf{b})\,\mathbf{H}$$

We are being sloppy at this stage by not distinguishing carefully between true values $\mathbf{b} = \boldsymbol{\beta}$ and estimated values $\mathbf{b} = \widehat{\boldsymbol{\beta}}$, but we will clean that up in a technical note.

Rearranging the above formula yields

$$\text{Var}(\mathbf{b}) \approx \mathbf{H}^{-1}\,\text{Var}(\mathbf{g})\,\mathbf{H}^{-1} \tag{3}$$

We showed that $\text{Var}(\mathbf{g}) = -E(\mathbf{H})$, so if we could prove that $E(\mathbf{H})$ is approximately $\mathbf{H}$ at the observed $\mathbf{X}$, then we would have

$$\text{Var}(\mathbf{b}) = \mathbf{H}^{-1}(-\mathbf{H})\mathbf{H}^{-1} = -\mathbf{H}^{-1} \tag{4}$$

which is what we wanted to show.

Below are two technical notes, the first providing a proof of the Lemma and the second repeating the casual proof above more formally. Also see Stuart and Ord (1991, 649–706).

❑ Technical Note

Proof of the Lemma $E(D\ln L) = \mathbf{0}$ and $-E(D^2 \ln L) = E((D\ln L)^2)$:

Note that $\int_{\mathbf{X}} L(\mathbf{b};\mathbf{X})\,d\mathbf{X} = 1$ since L is a density. Thus $D(\int_{\mathbf{X}} L\,d\mathbf{X}) = \mathbf{0}$. The standard line at this point is that "under appropriate regularity conditions", we can move the derivative (remember $D = \partial/\partial\mathbf{b}$) under the integral sign: $\int_{\mathbf{X}} DL\,d\mathbf{X} = \mathbf{0}$.

One might think that these regularity conditions are of no consequence for practical problems, but one of the conditions is that the range of $\mathbf{X}$ does not depend on $\mathbf{b}$. If it does, then all the following likelihood theory falls apart and the following estimation techniques will not work. When the range of $\mathbf{X}$ depends on $\mathbf{b}$, you have to start from scratch.

In any case,

$$\begin{aligned}\mathbf{0} = \int_{\mathbf{X}} DL\,d\mathbf{X} &= \int_{\mathbf{X}} (1/L)\,DL\,L\,d\mathbf{X} \\ &= \int_{\mathbf{X}} (D\ln L)\,L\,d\mathbf{X} \\ &= E(D\ln L)\end{aligned}$$

which concludes the first part of the proof. Taking the derivative of $\int_{\mathbf{X}}(D\ln L)\,L\,d\mathbf{X} = \mathbf{0}$ yields

$$\begin{aligned}D\int_{\mathbf{X}} (D\ln L)\,L\,d\mathbf{X} &= \int_{\mathbf{X}} D[(D\ln L)\,L]\,d\mathbf{X} \\ &= \int_{\mathbf{X}} [(D^2\ln L)L + (D\ln L)\,DL]\,d\mathbf{X} \\ &= \int_{\mathbf{X}} [(D^2\ln L) + (D\ln L)^2]\,L\,d\mathbf{X} \\ &= E(D^2\ln L) + E(\,(D\ln L)^2\,) = \mathbf{0}\end{aligned}$$

and thus $-E(D^2\ln L) = E(\,(D\ln L)^2\,)$, which concludes the second part of the proof.

❑

❑ Technical Note

Outside of the technical notes, we provided a casual proof that the variance of $\mathbf{b}$ is $-\mathbf{H}^{-1}$. We did not carefully distinguish, however, between $\widehat{\boldsymbol\beta}$, the estimates we obtain, and $\boldsymbol\beta$, the true values. What we really want to show is that our *estimator* of $\boldsymbol\beta$ has variance $-\mathbf{H}^{-1}$. There are two important consequences of being more careful:

1. All results are asymptotic, meaning that in finite samples, $\widehat{\boldsymbol\beta}$ may be a biased estimate of $\boldsymbol\beta$.

2. The variance estimate $-\mathbf{H}^{-1}$ is guaranteed to be the variance of $\widehat{\boldsymbol{\beta}}$ only when $\widehat{\boldsymbol{\beta}}$ becomes very close to $\boldsymbol{\beta}$. This has implications for hypothesis testing. In linear regression, by distinction, we test that a particular coefficient b_i is zero using the estimated variance, and that variance is equally applicable for both β_i and 0. That is not the case in general for maximum likelihood estimation. Under the null hypothesis $b_i = 0$, we need to know $\text{Var}(\widehat{\boldsymbol{\beta}})$ at the hypothesized value and that variance could be different from $-\mathbf{H}^{-1}$. (Aside: That is why likelihood-ratio tests are better than using Wald tests, which are based on $-\mathbf{H}^{-1}$; we will discuss this more later.)

Let $\boldsymbol{\beta}$ be the true value of the parameter vector we are trying to estimate. Let $\widehat{\boldsymbol{\beta}}$ be the maximum likelihood estimator of $\boldsymbol{\beta}$; that is, $\ln L(\mathbf{b};\mathbf{X})$ attains its maximum value at $\mathbf{b} = \widehat{\boldsymbol{\beta}}$.

$\mathbf{g}(\mathbf{b};\mathbf{X})$ is the score function and this time, to emphasize that $\mathbf{g}$ is a function of $\mathbf{b}$, we will write $\mathbf{g}(\mathbf{b})$. Likewise, $\mathbf{H}$ is also a function of $\mathbf{b}$ and we will write $\mathbf{H}(\mathbf{b})$.

Consistency of $\widehat{\boldsymbol{\beta}}$

By the Mean Value Theorem there exists a $\mathbf{b}^*$ between $\widehat{\boldsymbol{\beta}}$ and $\boldsymbol{\beta}$ such that $\mathbf{g}(\widehat{\boldsymbol{\beta}}) - \mathbf{g}(\boldsymbol{\beta}) = D\mathbf{g}(\mathbf{b}^*)\,(\widehat{\boldsymbol{\beta}} - \boldsymbol{\beta})$. Since $\widehat{\boldsymbol{\beta}}$ corresponds to the maximum of $\ln L$ and $\mathbf{g} = D\ln L$, therefore $\mathbf{g}(\widehat{\boldsymbol{\beta}}) = \mathbf{0}$. Since $D\mathbf{g} = \mathbf{H}$, we have

$$-\mathbf{g}(\boldsymbol{\beta}) = \mathbf{H}(\mathbf{b}^*)\,(\widehat{\boldsymbol{\beta}} - \boldsymbol{\beta}) \tag{5}$$

Note that

$$\mathbf{g}(\boldsymbol{\beta}) = \frac{\partial \ln \ell(\boldsymbol{\beta};\mathbf{x}_1)}{\partial \mathbf{b}} + \cdots + \frac{\partial \ln \ell(\boldsymbol{\beta};\mathbf{x}_N)}{\partial \mathbf{b}}$$

so that $\mathbf{g}(\boldsymbol{\beta})$ is the sum of N *i.i.d.* random variables. Thus, by the Strong Law of Large Numbers, it converges to its expectation:

$$\lim_{N\to\infty} \mathbf{g}(\boldsymbol{\beta}) = E\big(\mathbf{g}(\boldsymbol{\beta})\big) = \mathbf{0}$$

Since $\mathbf{H}(\mathbf{b}^*)$ is (asymptotically) nonsingular (we do not show this), the above and Equation 5 imply that

$$\lim_{N\to\infty} \widehat{\boldsymbol{\beta}} - \boldsymbol{\beta} = 0 \qquad \text{and thus} \qquad \lim_{N\to\infty} \widehat{\boldsymbol{\beta}} = \boldsymbol{\beta}$$

Thus $\widehat{\boldsymbol{\beta}}$ is a consistent estimator of $\boldsymbol{\beta}$.

Proof that $\widehat{\boldsymbol{\beta}}$ *is asymptotically normal with asymptotic variance* $-\mathbf{H}(\widehat{\boldsymbol{\beta}})^{-1}$

Let us rearrange Equation 5 to produce

$$\frac{\widehat{\boldsymbol{\beta}} - \boldsymbol{\beta}}{\sqrt{-\mathbf{H}(\mathbf{b}^*)^{-1}}} = \frac{\mathbf{g}(\boldsymbol{\beta})}{\sqrt{-\mathbf{H}(\mathbf{b}^*)}} \tag{6}$$

Note that what we have written above is not right for matrices, so let us suppose that $\mathbf{b}$ consists of only a single parameter. To do the proof correctly for multidimensional matrices involves quadratic forms and χ^2 distributions, but otherwise the proof is similar to what follows.

Since $\mathbf{b}^*$ is between $\widehat{\boldsymbol{\beta}}$ and $\boldsymbol{\beta}$ and $(\widehat{\boldsymbol{\beta}} - \boldsymbol{\beta}) \to 0$, thus $\mathbf{H}(\mathbf{b}^*)$, $\mathbf{H}(\boldsymbol{\beta})$, and $\mathbf{H}(\widehat{\boldsymbol{\beta}})$ are all asymptotically the same. Since $\mathbf{H}(\boldsymbol{\beta})$ is also the sum of *i.i.d.* random variables, by the Strong Law of Large Numbers they are all asymptotically equal to $E(\mathbf{H}(\boldsymbol{\beta}))$.

Recall Equation 2, $\text{Var}(\mathbf{g}(\boldsymbol{\beta})) = -E(\mathbf{H}(\boldsymbol{\beta}))$, and recall by our Lemma, $E(\mathbf{g}(\boldsymbol{\beta})) = 0$. Thus, the right-hand side of Equation 6 is asymptotically

$$\frac{\mathbf{g}(\boldsymbol{\beta}) - E(\mathbf{g}(\boldsymbol{\beta}))}{\sqrt{\text{Var}(\mathbf{g}(\boldsymbol{\beta}))}}$$

Since $\mathbf{g}(\boldsymbol{\beta})$ is the sum of N *i.i.d.* random variables, by the Central Limit Theorem, the above is asymptotically distributed $N(0,1)$. Hence the left-hand side of Equation 6 is also asymptotically distributed $N(0,1)$.

Since $\mathbf{H}(\mathbf{b}^*)$ was asymptotically the same as $\mathbf{H}(\widehat{\boldsymbol{\beta}})$, we write our final result:

$$\frac{\widehat{\boldsymbol{\beta}} - \boldsymbol{\beta}}{\sqrt{-\mathbf{H}(\widehat{\boldsymbol{\beta}})^{-1}}}$$

is asymptotically distributed $N(0,1)$ with $-\mathbf{H}(\widehat{\boldsymbol{\beta}})^{-1}$ the asymptotic variance of $\widehat{\boldsymbol{\beta}}$.

Q.E.D.

❑

1.3.1 All results are asymptotic

The first important consequence of the full proof is that all results are asymptotic:

1. $\widehat{\boldsymbol{\beta}} \to \boldsymbol{\beta}$ is guaranteed only as $N \to \infty$.
2. It is not true in general that $E(\widehat{\boldsymbol{\beta}}) = \boldsymbol{\beta}$ at finite N. $\widehat{\boldsymbol{\beta}}$ may be biased, and the bias may be significant at small N.
3. The variance of $\widehat{\boldsymbol{\beta}}$ is $-\mathbf{H}(\widehat{\boldsymbol{\beta}})^{-1}$ asymptotically. At finite N, we are not sure how good this variance estimate is.

Everyone knows this and knows never to estimate a maximum likelihood model with only a handful of observations.

1.3.2 The variance estimate may not be relevant for hypothesis tests

The variance estimate $-\mathbf{H}^{-1}$ may not be what you want for conducting hypothesis tests.

$\text{Var}(\widehat{\boldsymbol{\beta}})$ is $-\mathbf{H}^{-1}$ asymptotically for the true value $\boldsymbol{\beta}$ of $\mathbf{b}$ only. For hypothesis tests, we do not care about $\text{Var}(\widehat{\boldsymbol{\beta}})$ for the true value of $\mathbf{b}$. We need to know $\text{Var}(\widehat{\boldsymbol{\beta}})$ for the hypothesized value $\mathbf{b}_0$ of $\mathbf{b}$.

This may sound arcane but, if so, that is because you are too familiar with linear regression. In a linear regression model one can test that a particular coefficient b_i is, say, zero using the estimated variance because that variance is equally applicable for both β_i and 0. That is not necessarily the case in maximum likelihood estimation.

The idea behind hypothesis testing is that you determine the distribution of the estimator $\widehat{\boldsymbol{\beta}}$ when $\mathbf{b} = \mathbf{b}_0$ and you compute the p-value from that distribution. If you use the $-\mathbf{H}^{-1}$ variance estimates instead, it can make a little difference or a lot, but it will make more difference the farther $\mathbf{b}_0$ is away from $\boldsymbol{\beta}$.

Another way of thinking about this is that the estimate of variance $-\mathbf{H}^{-1}$ is obtained by taking a quadratic approximation to the log-likelihood function. The approximation is obtained by using a second-order Taylor series expansion:

$$\begin{aligned}\ln L(\mathbf{b}) &\approx \ln L(\widehat{\boldsymbol{\beta}}) + \mathbf{g}(\widehat{\boldsymbol{\beta}}) \cdot (\mathbf{b} - \widehat{\boldsymbol{\beta}}) + (\mathbf{b} - \widehat{\boldsymbol{\beta}})'\mathbf{H}(\mathbf{b} - \widehat{\boldsymbol{\beta}}) \\ &= \ln L(\widehat{\boldsymbol{\beta}}) + (\mathbf{b} - \widehat{\boldsymbol{\beta}})'\mathbf{H}(\mathbf{b} - \widehat{\boldsymbol{\beta}})\end{aligned}$$

The approximation is centered on $\mathbf{b} = \widehat{\boldsymbol{\beta}}$. If the log-likelihood function really is quadratic, then the quadratic fit is perfect even far from $\widehat{\boldsymbol{\beta}}$ and $-\mathbf{H}^{-1}$ is exactly the "right" variance estimate regardless of the hypothesis being tested. If, on the other hand, the log-likelihood function deviates from that, the approximation becomes poor.

1.3.3 Likelihood-ratio tests and Wald tests

A Wald test is a test of the coefficients based on the estimated variance $-\mathbf{H}^{-1}$. Stata's `test` command performs Wald tests.

Likelihood-ratio tests are based on comparing the heights of the likelihood function at $\widehat{\boldsymbol{\beta}}$ and $\mathbf{b}_0$. Stata's `lrtest` command performs likelihood-ratio tests.

The latter are widely viewed as better than the former because, as we just explained, the Wald test employs the "wrong" variance estimate.

The likelihood ratio is defined as

$$\text{LR} = \frac{\max_{\mathbf{b}=\mathbf{b}_0} L(\mathbf{b};\mathbf{X})}{\max_{\mathbf{b}} L(\mathbf{b};\mathbf{X})} = \frac{\max_{\mathbf{b}=\mathbf{b}_0} L(\mathbf{b};\mathbf{X})}{L(\widehat{\boldsymbol{\beta}};\mathbf{X})}$$

$\mathbf{b} = \mathbf{b}_0$ may be a simple hypothesis—all values of $\mathbf{b}$ are hypothesized to be some set of values such as $\mathbf{b}_0 = (0, 0, \dots, 0)$—or $\mathbf{b} = \mathbf{b}_0$ may be a composite hypothesis—only some values of $\mathbf{b}$ are hypothesized such as $\mathbf{b}_0 = (0, ?, ?, \dots, ?)$, where ? means that the value can be anything.

In general, we can write $\mathbf{b}_0 = (\mathbf{b}_r, \mathbf{b}_u)$, where $\mathbf{b}_r$ is fixed and $\mathbf{b}_u$ is not. Thus,

$$\max_{\mathbf{b}=\mathbf{b}_0} L(\mathbf{b};\mathbf{X}) = \max_{\mathbf{b}_u} L(\mathbf{b}_r, \mathbf{b}_u;\mathbf{X}) = L(\mathbf{b}_r, \widehat{\boldsymbol{\beta}}_u;\mathbf{X})$$

and the likelihood ratio becomes $\text{LR} = L(\mathbf{b}_r, \widehat{\boldsymbol{\beta}}_u;\mathbf{X})/L(\widehat{\boldsymbol{\beta}};\mathbf{X})$.

It can be shown that $-2\ln(\text{LR})$ has (asymptotically) a chi-squared distribution with r degrees of freedom where r is the dimension of $\mathbf{b}_r$. One important note about the theory is that it holds only for small to moderate r. If r is very large (say $r > 100$), the chi-squared distribution result becomes questionable.

Note that to compute the LR, we must do two maximizations: one to get $\widehat{\boldsymbol{\beta}}$ and another to compute $\widehat{\boldsymbol{\beta}}_u$. Hence LR tests can be time consuming because you must do an "extra" maximization for each hypothesis test.

The Wald test, on the other hand, simply uses $\text{Var}(\widehat{\boldsymbol{\beta}})$ estimated assuming $\mathbf{b} = \boldsymbol{\beta}$; i.e., it uses $-\mathbf{H}^{-1}$ for the variance. For a linear hypothesis $\mathbf{Rb} = \mathbf{r}$, it computes the Wald test statistic

$$W = (\mathbf{R}\widehat{\boldsymbol{\beta}} - \mathbf{r})'(\mathbf{R}\text{Var}(\widehat{\boldsymbol{\beta}})\mathbf{R}')^{-1}(\mathbf{R}\widehat{\boldsymbol{\beta}} - \mathbf{r})$$

which, assuming normality, has a chi-squared distribution with r degrees of freedom. Obviously, this is an easy computation to make once we have $\text{Var}(\widehat{\boldsymbol{\beta}})$. The problem with the Wald test is that we want $\text{Var}(\widehat{\boldsymbol{\beta}})$ computed when $\mathbf{b} = \mathbf{b}_0$, not $\mathbf{b} = \boldsymbol{\beta}$.

If you really care about that p-value, use the likelihood-ratio test.

1.3.4 Robust variance estimates

Our result that $\text{Var}(\widehat{\boldsymbol{\beta}})$ is asymptotically $-\mathbf{H}^{-1}$ hinges on the Lemma we first proved. It was in proving the Lemma and only in proving the Lemma that we assumed the likelihood function $L(\mathbf{b};\mathbf{X})$ was the density function for $\mathbf{X}$. If $L(\mathbf{b};\mathbf{X})$ is not the true density function, our Lemma does not apply and all our subsequent results do not necessarily hold. In practice, various work has shown that the maximum likelihood estimator and its variance estimate still work reasonably well if $L(\mathbf{b};\mathbf{X})$ is a little off from the true density function. For example, if the true density is probit and you estimate a logit model (or vice versa), then the results are still fine.

Alternatively, one can derive a variance estimator that does not need $L(\mathbf{b};\mathbf{X})$ to be the density function for $\mathbf{X}$. This is the robust variance estimator which is implemented in many Stata estimation commands and in the survey (`svy`) commands.

The robust variance estimator was first published, we believe, by Peter Huber (a mathematical statistician) in 1967 in a conference proceedings (Huber 1967). Survey statisticians were thinking about the same things around this time, at least for linear regression. In the 1970s, the survey statisticians wrote up their work, including Kish and Frankel (1974), Fuller (1975), and others; all of which was summarized and generalized in an excellent paper by Binder (1983). White, an economist, independently derived the estimator and published it in 1980 (linear regression) and 1982 (MLEs) in the economics literature (White 1980, 1982). Many others have extended its development including Gail, Tan, and Piantadosi (1988), Kent (1982), Royall (1986), and Lin and Wei (1989).

The robust variance estimator is called different things by different people. At Stata, we originally called it the Huber variance estimator (Bill Rogers, who first implemented it here, was a student of Huber's). Some people call it the sandwich estimator. Survey statisticians call it the Taylor-series linearization method, or linearization method, or design-based variance estimate. Economists often call it the White estimator. Statisticians often refer to it as the empirical variance estimator.

In any case, it is the same estimator.

We will just sketch the derivation here.

The starting point is Equation 3 of Section 1.3: $\text{Var}(\mathbf{b}) \approx \mathbf{H}^{-1}\text{Var}(\mathbf{g})\mathbf{H}^{-1}$. We want to evaluate the formula at $\mathbf{b} = \widehat{\boldsymbol{\beta}}$, so we will write it

$$\text{Var}(\widehat{\boldsymbol{\beta}}) \approx \mathbf{H}(\widehat{\boldsymbol{\beta}})^{-1}\text{Var}(\mathbf{g}(\widehat{\boldsymbol{\beta}}))\mathbf{H}(\widehat{\boldsymbol{\beta}})^{-1} \tag{3'}$$

It is due to this starting point that some people refer to robust variance estimates as the Taylor-series linearization method. Equation 3 was obtained using the delta method, which is based on a first-order (linear term only) Taylor series.

It is the next step that causes other people to refer to it as the empirical variance estimate. Note that

$$\mathbf{g}(\widehat{\boldsymbol{\beta}};\mathbf{X}) = \frac{\partial \ln \ell(\widehat{\boldsymbol{\beta}};\mathbf{x}_1)}{\partial \mathbf{b}} + \cdots + \frac{\partial \ln \ell(\widehat{\boldsymbol{\beta}};\mathbf{x}_N)}{\partial \mathbf{b}}$$

which is to say, $\mathbf{g}$ is the sum of N *i.i.d.* random variables. What is a good estimator for the variance of the sum of N *i.i.d.* random variables? Do not let the fact that we are summing evaluations of functions fool you. A good estimator for the variance of the mean of N *i.i.d.* random variates z_1, z_2, ..., z_N is, as everyone knows,

$$\frac{s^2}{N} = \frac{1}{N(N-1)}\sum_{j=1}^{N}(z_j - \overline{z})^2$$

Therefore a good estimator for the variance of the total is simply N^2 times it:

$$Ns^2 = \frac{N}{N-1}\sum_{j=1}^{N}(z_j - \overline{z})^2$$

Thus, we simply use $z_j = \mathbf{g}(\widehat{\boldsymbol\beta}; \mathbf{x}_j)$ in this formula to obtain an estimate of $\text{Var}(\mathbf{g}(\widehat{\boldsymbol\beta}))$ (and we substitute $\overline{z} = 0$ because $\mathbf{g}(\widehat{\boldsymbol\beta}; \mathbf{X}) = \mathbf{0}$). If we then plug this empirical estimate of $\text{Var}(\mathbf{g}(\widehat{\boldsymbol\beta}))$ into $\text{Var}(\widehat{\boldsymbol\beta}) \approx \mathbf{H}(\widehat{\boldsymbol\beta})^{-1}\text{Var}(\mathbf{g}(\widehat{\boldsymbol\beta}))\mathbf{H}(\widehat{\boldsymbol\beta})^{-1}$, we obtain

$$\text{Var}(\widehat{\boldsymbol\beta}) \approx \mathbf{H}(\widehat{\boldsymbol\beta})^{-1}\left(\frac{N}{N-1}\sum_{j=1}^{N}\mathbf{g}_j\,\mathbf{g}_j'\right)\mathbf{H}(\widehat{\boldsymbol\beta})^{-1}$$

This is the robust variance estimator. It is robust for the same reasons that the estimator for the variance of a mean is robust; note our lack of assumptions about $\ln \ell$ and hence L.

The estimator for the variance of the total of $\mathbf{g}_j$ relies only on the fact that we have simple random sampling. Thus, this is the essential assumption of the robust variance estimator: The observations are independent selections from the same population.

For cluster sampling, we merely change our estimator for the variance of the total $\mathbf{g}_j$ to reflect this sampling scheme. Consider super-observations made up of the sum of $\mathbf{g}_j$ for a cluster: These super-observations are independent and the above formulas hold with $\mathbf{g}_j$ replaced by the cluster sums. See [R] **_robust** and [R] **svymean** in the *Stata Reference Manual* for more details.

1.4 The maximization problem

Given the problem

$$\max_{\mathbf{b}} \quad L(\mathbf{b}; \mathbf{X})$$

how do we obtain the solution? The way we would solve this analytically is by taking derivatives and setting them to zero:

$$\text{Solve for } \mathbf{b}: \qquad \frac{\partial \ln L()}{\partial \mathbf{b}} = \mathbf{0}$$

As an overly simplified example, if $\ln L() = b^2 + b$, then

$$\frac{d\ln L(b)}{db} = 2b + 1 = 0 \qquad \Rightarrow \qquad b = -1/2$$

In general, however, $\partial \ln L/\partial \mathbf{b} = \mathbf{0}$ is too complicated to admit an analytic solution and we are forced to seek a numerical one. We start in the same way

$$\text{Solve for } \mathbf{b}: \qquad \frac{\partial \ln L()}{\partial \mathbf{b}} = \mathbf{0}$$

and, if we write $\mathbf{g}(\mathbf{b}; \mathbf{X})$ for $\partial \ln L()/\partial \mathbf{b}$, our problem is

$$\text{Solve for } \mathbf{b}: \qquad \mathbf{g}(\mathbf{b}; \mathbf{X}) = \mathbf{0}$$

Note that we have converted the maximization problem into a root-finding problem. Estimates $\mathbf{b}$ are obtained by finding the roots of $\mathbf{g}(\mathbf{b}; \mathbf{X})$. Computer optimizers (the generic name for maximizers or minimizers) are, at heart, root finders.

1.4.1 Numerical root finding

For one dimension (b and $g(b)$ both scalars), one method of finding roots is Newton's Method. This is an iterative technique. You have a guess b_0 (called the initial value) and you update that as follows:

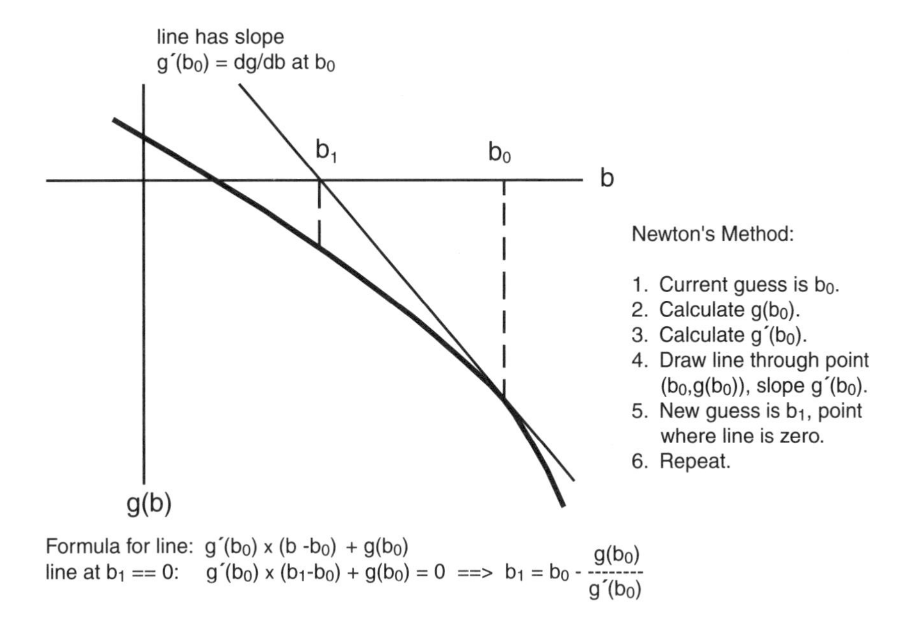

Now you have a new value b_1 and you can repeat the process,

$$b_2 = b_1 - \frac{g(b_1)}{g'(b_1)}$$

and so on. The sequence is not guaranteed to converge, but it generally does.

Newton's Method for finding roots can be converted to finding minimums and maximums of $f(b)$ by substituting $f'()$ for $g()$ and $f''()$ for $g'()$:

To find b such that $f(b)$ is maximized,

1. Start with guess b_0.
2. Calculate new guess $b_1 = b_0 - f'(b_0)/f''(b_0)$.
3. Repeat.

We can generalize this to allow b to be a vector $\mathbf{b}$:

To find vector $\mathbf{b}$ such that $f(\mathbf{b})$ is maximized,

1. Start with guess $\mathbf{b}_0$.
2. Calculate new guess $\mathbf{b}_1 = \mathbf{b}_0 - \mathbf{g}\mathbf{H}^{-1}$, where $\mathbf{g}$ is the gradient vector and $\mathbf{H}$ the matrix of second derivatives. (In the future, we will write this $\mathbf{b}_1 = \mathbf{b}_0 + \mathbf{g}(-\mathbf{H})^{-1}$ because the matrix $-\mathbf{H}$ is positive definite and calculating the inverse of a positive-definite matrix is easier.)
3. Repeat.

This is the Newton–Raphson algorithm and is, in fact, the method Stata uses, as do most other statistical packages. Actually, there are a few details that are commonly changed and, in fact, any package that claims to use Newton–Raphson is probably oversimplifying the story. These details are changed because

1. Calculating $f(\mathbf{b}) = \ln L(\mathbf{b}; \mathbf{X})$ is computationally expensive.
2. Calculating $\mathbf{g}(\mathbf{b}) = \partial \ln L()/\partial \mathbf{b}$ is even more computationally expensive.
3. Calculating $\mathbf{H}(\mathbf{b}) = \partial^2 \ln L()/\partial \mathbf{b} \partial \mathbf{b}'$ is even more computationally expensive than that.

Relatively speaking, we can calculate $f(\mathbf{b})$ many times for the same computer time required to calculate $\mathbf{H}(\mathbf{b})$ just once. This leads to separating the direction calculation from the stepsize:

To find vector $\mathbf{b}$ such that $f(\mathbf{b})$ is maximized,

1. Start with a guess $\mathbf{b}_0$.
2. Calculate a direction vector $\mathbf{d} = \mathbf{g}(-\mathbf{H})^{-1}$.
3. Calculate a new guess $\mathbf{b}_1 = \mathbf{b}_0 + s\mathbf{d}$, where s is a scalar.
   a. For instance, start with $s = 1$.
   b. If $f(\mathbf{b}_0 + \mathbf{d}) > f(\mathbf{b}_0)$, try $s = 2$. If $f(\mathbf{b}_0 + 2\mathbf{d}) > f(\mathbf{b}_0 + \mathbf{d})$, try $s = 3$ or even $s = 4$, and so on.
   c. If $f(\mathbf{b}_0 + \mathbf{d}) \leq f(\mathbf{b}_0)$, back up and try $s = .5$ or $s = .25$, etc.
4. Repeat.

The way that s is obtained varies from package to package but the idea is simple enough: go in the direction $\mathbf{d} = \mathbf{g}(-\mathbf{H})^{-1}$ as far as one possibly can before spending computer cycles to compute a new direction. This all goes under the rubric "stepsize calculation".

The other important detail goes under the rubric "nonconcavity" and all that means is that $(-\mathbf{H})^{-1}$ does not exist—$\mathbf{H}$ is singular—and the direction calculation $\mathbf{d} = \mathbf{g}(-\mathbf{H})^{-1}$ is impossible. It is worth thinking about this case in the single-dimension case: In the maximization problem, it means that the second derivative of $f()$ is zero. In the root-finding problem, it means that the derivative of $g()$ is zero. The gradient is flat and so there is no root:

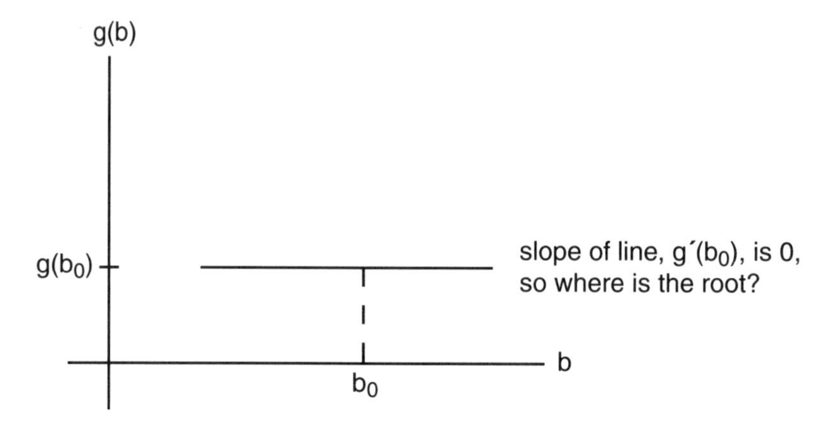

Remember, all you really know is that $g'()$ evaluated at b_0 is 0. Even though we have drawn a line, you do not really know that $g'(b) = 0$ for all b. In fact, you are reasonably certain that "$g'(b) = 0$ for all b" is not true because $g()$ is the gradient of a likelihood function and the likelihood certainly cannot be increasing or decreasing without bound. Probably b_0 is just a poor guess and far from the heart of the function.

The line we have drawn is how you would usually project to find the new guess b_1—it goes through the point $(b_0, g(b_0))$ and has slope $g'(b_0)$. We graphed it to emphasize why the usual rule is no help in this case.

So pretend that you are the optimizer charged with finding the maximum of $f(b)$, b_0 is the current guess, and $g'(b_0) = 0$. What is your next guess? Put that aside. Is your next guess b_1 to the left or right of b_0?

This is where the knowledge that this is a maximization problem rather than merely a root-finding problem is useful. If you told us just to find the roots, we would have no reason to select $b_1 > b_0$ or $b_1 < b_0$; we would have to try both. But if you tell us this is a maximization problem, we observe that in the graph drawn, $g(b_0) > 0$, so increasing b increases the value of $f()$. We would try something to the right of b_0. (If $g(b_0) < 0$, we would try something to the left of b_0 and, if this were a minimization rather than a maximization problem, we would reverse the rules.)

The important point is that we are not without any idea of what direction to go merely because $g'(b_0) = 0$.

When we generalize to allow b to be a vector, our choices proliferate. Corresponding to our scalar go-with-the-sign-of-$g()$ rule, the vector equivalent is to go-with-the-vector-$\mathbf{g}()$ rule, which amounts to treating the noninvertible matrix $\mathbf{H}$ as if $-\mathbf{H}^{-1} = \mathbf{I}$ since our general rule is direction $\mathbf{d} = \mathbf{g}(-\mathbf{H})^{-1}$. That is called steepest ascent. It is called ascent because, just as in the scalar case, we choose the direction based on sign of the gradient and knowledge that we want to maximize $f()$. It is called steepest because, mathematically, it works out that among all directions we could go with $f()$ increasing, the direction $\mathbf{g}()$ is steepest. (By the way, the steepest-ascent rule is often called the steepest-descent rule because most of the numerical optimization literature is written in terms of function minimization rather than maximization.)

This rule can be improved upon. Rather than treating $-(\mathbf{H})^{-1}$ as $\mathbf{I}$, one can add a constant to the diagonal elements of $-\mathbf{H}$ until $-\mathbf{H}$ is invertible. This is like mixing in a little steepest-ascent to the otherwise "best" direction and is called the modified Marquardt algorithm (Marquardt, 1963).

Stata follows one of two variations on this scheme. In the first variation, rather than adding a constant c to the diagonal terms, `ml` adds a proportional term $c|h_{ii}|$. `ml` then renormalizes the resulting matrix so that it has the original trace (thus attempting to keep the scale of the matrix the same).

If you specify `ml`'s `difficult` option, it does more work. In that case `ml` computes the eigenvalues of $-\mathbf{H}$ and then, for the part of the orthogonal subspace where the eigenvalues are negative or small positive numbers, uses steepest ascent and in the other subspace uses a regular Newton–Raphson step.

1.4.2 Numerical maximization

To summarize, to find $\mathbf{b}$ such that $f(\mathbf{b})$ is maximized,

1. Start with a guess $\mathbf{b}_0$.
2. Calculate a direction vector $\mathbf{d} = \mathbf{g}(-\mathbf{H})^{-1}$ if you can. If $-\mathbf{H}$ is not invertible, substitute for $-\mathbf{H}$ the matrix $\mathbf{I}$, or $-\mathbf{H} + c\mathbf{I}$ for some small, positive scalar c, or some other matrix related to $-\mathbf{H}$.

3. Calculate a new guess $\mathbf{b}_1 = \mathbf{b}_0 + s\mathbf{d}$, where s is a scalar.
   a. For instance, start with $s = 1$.
   b. If $f(\mathbf{b}_0 + \mathbf{d}) > f(\mathbf{b}_0)$, try $s = 2$. If $f(\mathbf{b}_0 + 2\mathbf{d}) > f(\mathbf{b}_0 + \mathbf{d})$, try $s = 3$ or even $s = 4$, and so on.
   c. If $f(\mathbf{b}_0 + \mathbf{d}) \leq f(\mathbf{b}_0)$, back up and try $s = .5$ or $s = .25$, etc.
4. Repeat.

If you have ever used Stata's `ml` command or any of Stata's preprogrammed maximum likelihood estimators, you have probably seen remnants of these rules.

When Stata says

```
Iteration #:  log likelihood = ...
```

That means it is at Step 2 and calculating a direction. The log likelihood reported is $f(\mathbf{b}_0)$.

Have you ever seen

```
Iteration #:  log likelihood = ...   (not concave)
```

That arose during Step 2. $-\mathbf{H}$ was not invertible, and Stata had to use something else to choose a direction. So what does it mean? If the message occurs in early iterations, you can safely ignore it. It just means that the current estimate of $\mathbf{b}$ is rather poor and that the matrix of second derivatives calculated at that point is noninvertible—corresponding in the scalar case to the gradient being flat.

In the multidimensional case, a noninvertible $-\mathbf{H}$ means that the likelihood function $f()$ locally has a ridge or a flat section or a saddle point. Real likelihood functions have few points like this and when such arise, it is typically because the gradient is very, but not perfectly, flat. Roundoff error makes the nearly flat gradient appear absolutely flat. In any case, the optimizer encountered the problem and worked around it.

If, on the other hand, this message were to arise at the final iteration, just before results are presented, it would be of great concern.

Have you ever seen Stata say

```
Iteration #:  log likelihood = ...   (backed up)
```

That arose in Step 3. Stata started with $s = 1$ and, with $s = 1$, it expected $f(\mathbf{b}_0 + s\mathbf{d}) = f(\mathbf{b}_0 + \mathbf{d}) > f(\mathbf{b}_0)$. That did not happen. Actually, Stata only mentions that it backed up if it halved the original stepsize six or more times. As with the not-concave message, this should not concern you unless it occurs at the last iteration, in which case the declaration of convergence is questionable. (You will learn to specify `ml`'s `gradient` option to see the gradient vector in cases such as this so you can determine whether its elements really are close to zero.)

1.4.3 Numerical derivatives

The maximizer that we have sketched has substantial analytic requirements. In addition to specifying $f(\mathbf{b}) = \ln L(\mathbf{b}; \mathbf{X})$, the routine needs to be able to calculate $\mathbf{g}(\mathbf{b}) = \partial f / \partial \mathbf{b}$ and $\mathbf{H}(\mathbf{b}) = \partial^2 f / \partial \mathbf{b} \partial \mathbf{b}'$. It needs the function, its first derivatives, and its second derivatives.

Stata provides facilities so that you do not actually have to program the derivative calculations, but really, Stata has only one optimizer. When you do not provide the derivatives, Stata calculates them numerically.

The definition of an analytic derivative is

$$f'(x_0) = \left. \frac{df}{dx} \right|_{x_0} = \lim_{h \to 0} \frac{f(x_0 + h) - f(x_0)}{h}$$

and that leads to a simple approximation formula,

$$f'(x_0) \text{ is approximately } \quad \frac{f(x_0+h)-f(x_0)}{h} \quad \text{for a suitable small but large enough } h$$

We have a lot to say about how Stata finds a suitably small but large enough h, and about variations on this formula, but put all that aside. The fact that Stata uses a formula like $[f(x_0+h)-f(x_0)]/h$ has an important implication for your program that calculates $f()$, the log-likelihood function.

If Stata chooses h just right, in general this formula will only be about half as accurate as the routine that calculates $f()$. Because of how Stata chooses h, it will not be less accurate than that and it might be more accurate. The loss of accuracy comes about because of the subtraction in the numerator.

Let us show you: consider calculating the derivative of $\exp(x)$ at $x = 2$. The true answer is $\exp(2)$ because $d(\exp(x))/dx = \exp(x)$. Let's try this formula with $h = 10^{-8}$ (h = 1e–8 in computer speak). We will carefully do this calculation with 16 digits of accuracy:

```
exp(2 + 1e-8)      =  7.389056172821211    (accuracy is 16 digits)
exp(2)             =  7.389056098930650    (accuracy is 16 digits)
---------------------------------------
difference         =   .000000073890561

exp(2)             =  7.389056098930650    (true answer)
difference / 1e-8  =  7.389056033701991    (approximation formula)
---------------------------------------
error                  .000000065228659    approximation is correct to
                      1 2345678            8 digits
```

The major source of error was introduced when we calculated the difference $\exp(2+1\text{e}{-8})-\exp(2)$.

```
exp(2 + 1e-8)      =  7.389056172821211    (accuracy is 16 digits)
exp(2)             =  7.389056098930650    (accuracy is 16 digits)
---------------------------------------
difference         =   .000000073890561
                               12345678    (accuracy is 8 digits)
```

Our full 16 digits of accuracy was lost since half the digits of $\exp(2+1\text{e}{-8})$ and $\exp(2)$ were in common and so canceled.

This is an unpleasant feature of numerical derivatives. Given how Stata chooses h, if you start with k digits of accuracy, you will fall to $k/2$ digits at worst. You can get lucky, the best case being when $f(x_0) = 0$ in which case you can get all k digits back, but that is ignoring the inaccuracy introduced by the algorithm itself.

Thus, as a programmer of likelihood functions, if you are going to use numerical derivatives, it is vitally important that you supply $\ln L()$ as accurately as you can. This means all your `generate` statements should explicitly specify `double`:

```
. gen double ... = ...
```

If you do not specify the `double`, Stata will store the result as `float`, which has about 7 digits of accuracy, meaning the numerical first derivatives will be accurate to only 3.5 digits. (Even if you specify `double`, your routine is probably not accurate to 16 digits; it is merely carrying around 16 digits and the roundoff error inherent in your algorithm probably means you will supply only 13 or 14 accurate digits and may be less. The minute you include one `generate` without the `double`, however, accuracy falls to 7 digits and then the roundoff error further reduces it from there.)

The issue in calculating numerical derivatives is choosing h. Stata does that for you. Just so that you are properly appreciative, let's consider that problem:

1. If Stata chooses a value for h that is too small, $f(x_0 + h)$ will numerically equal $f(x_0)$ and then the approximation to $f'(x_0)$ will be zero.
2. Values larger than that, but still too small, result in poor accuracy due to the numerical roundoff error associated with subtraction that we have just seen. If $f(x_0 + h)$ is very nearly equal $f(x_0)$, nearly all the digits are in common and the difference has few significant digits. For instance, say $f(x_0)$ and $f(x_0 + h)$ differ only in the 15th and 16th digits:

```
f(x0+h)          =  x.xxxxxxxxxxxxxab
f(x0)            =  x.xxxxxxxxxxxxxcd
--------------------------------------
difference       =  0.0000000000000ef   (2 digits of accuracy)
```

   Even worse, the 16-digit numbers shown might be accurate to only 14 digits themselves. In that case, the difference $f(x_0 + h) - f(x_0)$ would be purely noise.
3. On the other hand, if Stata chooses a value for h that is too large, the approximation formula itself will not estimate the derivative accurately.

Choosing the right value for h is of great importance. Above we considered numerically evaluating the derivative of $\exp(x)$ at $x = 2$. Below we calculate the error for a variety of h's:

```
. drop _all

. input h

              h
  1. 1e-20
  2. 1e-19
  3. 1e-18
(output omitted)
 19. 1e-2
 20. 1e-1
 21. 1
 22. end

. gen double approx = (exp(2+h)-exp(2))/h

. gen double error = exp(2) - approx

. gen double relerr = error/exp(2)

. list

              h      approx       error       relerr
  1.   1.00e-20           0   7.3890561            1
  2.   1.00e-19           0   7.3890561            1
  3.   1.00e-18           0   7.3890561            1
  4.   1.00e-17           0   7.3890561            1
  5.   1.00e-16           0   7.3890561            1
  6.   1.00e-15   6.2172489   1.1718072    .15858686
  7.   1.00e-14   7.5495167   -.1604606   -.02171598
  8.   1.00e-13   7.3807628    .0082933    .00112238
  9.   1.00e-12   7.3896445  -.00058838   -.00007963
 10.   1.00e-11   7.3890228   .00003334    4.512e-06
 11.   1.00e-10   7.3890582  -2.057e-06   -2.783e-07
 12.   1.00e-09   7.3890567  -5.878e-07   -7.956e-08
 13.   1.00e-08   7.3890561   2.032e-08    2.750e-09
 14.   1.00e-07   7.3890565  -3.636e-07   -4.920e-08
 15.   1.00e-06   7.3890598  -3.694e-06   -5.000e-07
 16.   1.00e-05    7.389093  -.00003695   -5.000e-06
 17.      .0001   7.3894256  -.00036947       -.00005
 18.       .001   7.3927519  -.00369576   -.00050017
 19.        .01   7.4261248  -.03706874   -.00501671
 20.         .1   7.7711381  -.38208204   -.05170918
 21.          1   12.696481  -5.3074247   -.71828183
```

In comparing these alternative h's, you should focus on the relerr column. The minimum relative error in this experiment is recorded in observation 13, $h = 1e{-}8$.

So, should h be 1e–8? Not in general. 1e–8 is a good choice for calculating the derivative of exp() at 2, but at some other location, a different value of h would be best. Even if you map out the entire exp() function, you will not have solved the problem. Change the function and you change the best value of h.

Nash (1990, 219) suggested that one choose h so that $x + h$ differs from x in at least half of its digits (the least significant ones). In particular, Nash suggested the formula

$$h(x) = \epsilon\,(|x| + \epsilon)$$

where ϵ is the square root of machine precision. Machine precision is roughly 1e–16 for double precision and therefore $h(2) \approx$ 2e–8. This way of dynamically adjusting h to the value of x works pretty well.

Nash's suggestion can be improved upon if one is willing to spend computer time. There are two issues that require balancing:

1. If the calculation were carried out in infinite precision, the smaller is h, the more accurately $[f(x + h) - f(x)]/h$ would approximate the derivative.
2. Taking into consideration finite precision, the closer are the values $f(x + h)$ and $f(x)$, the fewer are the significant digits in the calculation $f(x + h) - f(x)$.

Thus, we use a variation on Nash's suggestion and control the numerator:

$f(x + h)$ and $f(x)$ should differ in about half their digits

That is, we adopt the rule of setting h as small as possible subject to the constraint that $f(x+h) - f(x)$ will be calculated to at least half accuracy.

This is a computationally expensive rule to implement because, each time a numerical derivative is to be calculated, the program must search for the optimal value of h, meaning that $f(x + h)$ must be calculated for trial values. The payoff, however, is that `ml` is remarkably robust.

`ml` also uses a centered derivative calculation. Rather than using

$$f'(x) \approx \frac{f(x + h) - f(x)}{h}$$

`ml` uses

$$f'(x) \approx \frac{f(x + h/2) - f(x - h/2)}{h}$$

This also ups the computational time required but results in an important improvement, reducing the error (ignoring numerically induced error) from being $O(h)$ to $O(h^2)$.

1.4.4 Numerical second derivatives

Remember that Newton–Raphson not only requires $f'(x)$, it requires $f''(x)$, the second derivatives, too. We use the same method for calculating this as we use for calculating the first derivatives:

$$f''(x) \approx \frac{f'(x + h/2) - f'(x - h/2)}{h}$$

If you make the substitution for $f'(x)$, you obtain

$$f''(x) \approx \frac{f(x+h) - 2f(x) + f(x-h)}{h^2}$$

Think about the accuracy of this. You might first be tempted to reason that if $f()$ is calculated to 16 decimal places, $f'()$ will be accurate to $16/2$ decimal places at worst, assuming optimal h. If $f'()$ is accurate to 8 decimal places, $f''()$ will be accurate to at least $8/2$ decimal places, at worst. Thus, $f''()$ is accurate to at least 4 decimal places, at worst.

That is not right: you must look at the numerical calculation formula. The numerator can be rewritten

$$\big(f(x+h) - f(x)\big) + \big(f(x-h) - f(x)\big)$$

The first term $f(x+h) - f(x)$ can be calculated to half accuracy. The second term can similarly be calculated to the same accuracy if we assume $f(x+h)$ and $f(x-h)$ are of the same order of magnitude (which is binary magnitude for modern digital computers). Given that roughly half the digits have already been lost, the numerical summation of the two half-accurate results can be done without further loss of precision, at least assuming that they are of the same order of (binary) magnitude. The net result is that the numerator can be calculated nearly as accurately as the numerator for the first derivatives.

1.5 Monitoring convergence

When you perform maximum likelihood estimation, prior to presenting the estimates, Stata displays something that looks like this:

```
Iteration 0:  log likelihood =  -45.03321
Iteration 1:  log likelihood = -27.990675
Iteration 2:  log likelihood = -23.529093
Iteration 3:  log likelihood = -21.692004
Iteration 4:  log likelihood = -20.785625
Iteration 5:  log likelihood = -20.510315
Iteration 6:  log likelihood = -20.483776
Iteration 7:  log likelihood = -20.483495
```

Partly, Stata displays this for the entertainment value. Placing something on the screen periodically convinces you that your computer is still working.

The iteration log, however, contains reassuring information if you know how to read it. Now that you know how Stata's modified Newton–Raphson works—it is based on Newton's method for finding roots—it should be obvious to you that for smooth functions it may jump around at first but, once it gets close to the maximum (root), it should move smoothly toward it, taking smaller and smaller steps.

That is what happened above. Here is a graph of the iterations:

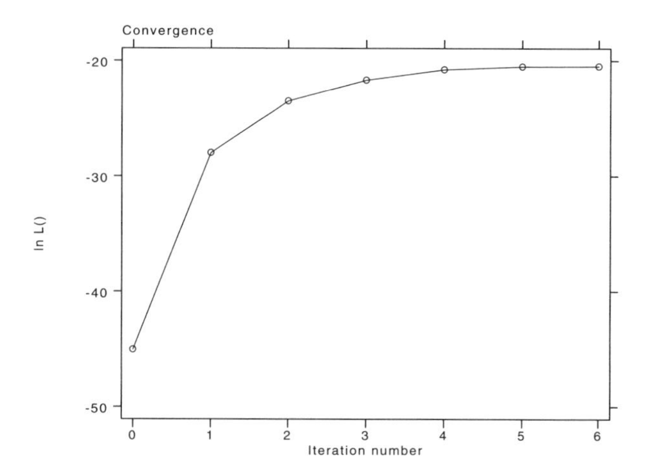

When we see this, we are reassured that the numerical routines have really converged to a maximum. Every step changes the log likelihood by less than the previous step.

It would not bother us too much if the early iterations jumped around as long as the later iterations followed the expected pattern:

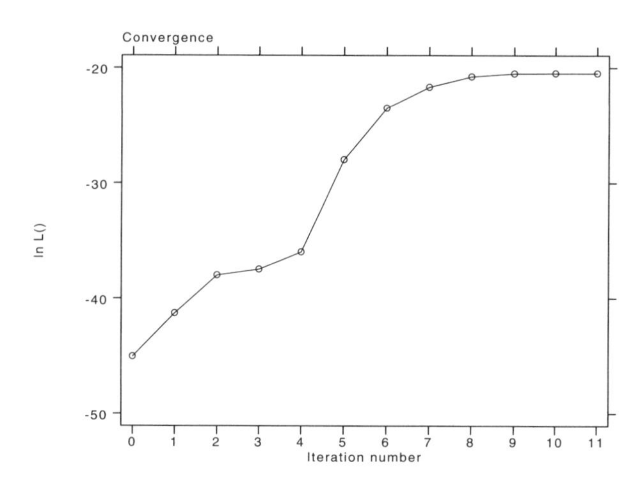

All this means is that, wherever we were at the 2nd through 4th iterations, the likelihood function flattened out.

This, however, would concern us

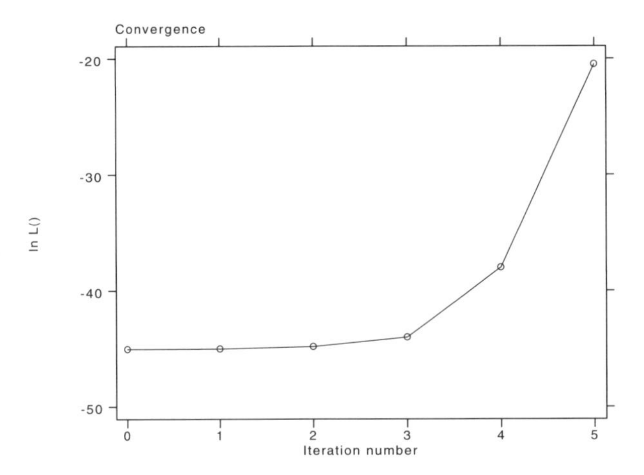

and so would this:

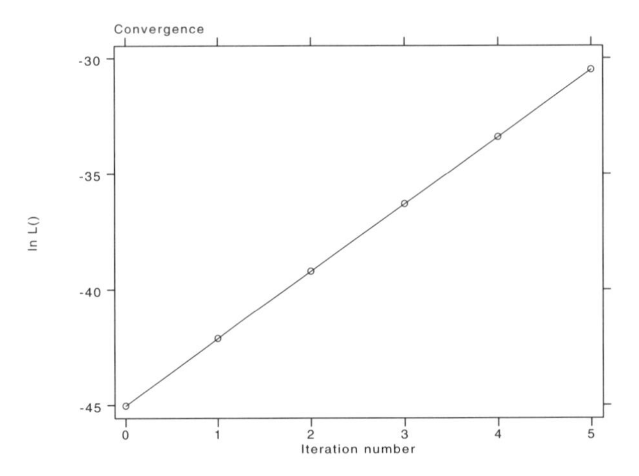

They would concern us because we are not seeing the expected slowing down of increments in the likelihood function as we get close to the solution.

In the last example, after thinking about the possibility of a programming bug and dismissing it, we would force more iterations and hope to see

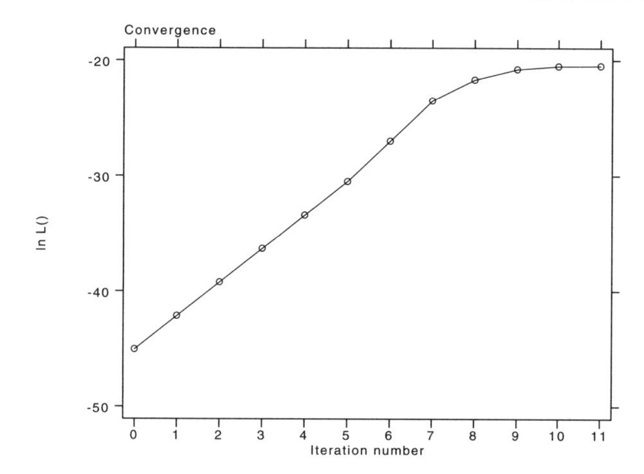

but we would not be shocked to see:

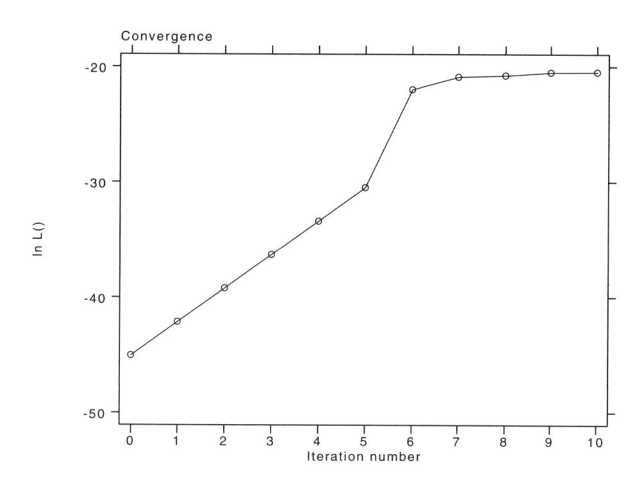

Either way, it would mean that we just had not done enough iterations previously.

You will learn that, after estimation, you can type **`ml graph`** to obtain graphs like the ones above. We suggest you use that feature.

Stata is reasonably conservative about declaring convergence and, in fact, it is not uncommon to see irritating evidence of that:

```
Iteration 0:  log likelihood =  -45.03321
Iteration 1:  log likelihood = -27.990675
Iteration 2:  log likelihood = -23.529093
Iteration 3:  log likelihood = -21.692004
Iteration 4:  log likelihood = -20.785625
Iteration 5:  log likelihood = -20.510315
Iteration 6:  log likelihood = -20.483776
Iteration 7:  log likelihood = -20.483495
Iteration 8:  log likelihood = -20.483495
```

Note that iterations 7 and 8 both present the same value (to the sixth decimal place) of the log-likelihood function. This extra step can be irritating when iterations take 10 or 15 seconds, but the equal value at the end is very reassuring. Stata does not always do that, but we like to see it the first time we code a likelihood function and, whatever Stata does, we invariably force a few extra steps in testing just for the reassurance.

1.6 References

Binder, D. A. 1983. On the variances of asymptotically normal estimators from complex surveys. *International Statistical Review* 51: 279–292.

Fuller, W. A. 1975. Regression analysis for sample survey. *Sankhyā, Series C* 37: 117–132.

Gail, M. H., W. Y. Tan, and S. Piantadosi. 1988. Tests for no treatment effect in randomized clinical trials. *Biometrika* 75: 57–64.

Huber, P. J. 1967. The behavior of maximum likelihood estimates under non-standard conditions. In *Proceedings of the Fifth Berkeley Symposium in Mathematical Statistics and Probability*. Berkeley, CA: University of California Press, 1, 221–233.

Kent, J. T. 1982. Robust properties of likelihood ratio tests. *Biometrika* 69: 19–27.

Kish, L. and M. R. Frankel. 1974. Inference from complex samples. *Journal of the Royal Statistical Society* B 36: 1–37.

Lin, D. Y. and L. J. Wei. 1989. The robust inference for the Cox proportional hazards model. *Journal of the American Statistical Association* 84: 1074–1078.

Marquardt, D. W. 1963. An algorithm for least squares estimation of nonlinear parameters. *Journal of the Society for Industrial and Applied Mathematics* 11: 431–441.

Nash, J. C. 1990. *Compact Numerical Methods for Computers: Linear Algebra and Function Minimisation.* 2d ed. New York: Adam Hilger.

Royall, R. M. 1986. Model robust confidence intervals using maximum likelihood estimators. *International Statistical Review* 54: 221–226.

Stuart, A. and J. K. Ord. 1991. *Kendall's Advanced Theory of Statistics, Vol. 2.* 5th ed. New York: Oxford University Press.

White, H. 1980. A heteroskedasticity-consistent covariance matrix estimator and a direct test for heteroskedasticity. *Econometrica* 48: 817–830.

——. 1982. Maximum likelihood estimation of misspecified models. *Econometrica* 50: 1–25.

2 Overview of ml

Contents

2.1 Basic syntax

`ml model` *method progname eq* [*eq* ...] [*weight*] [`if` *exp*] [`in` *range*]
    [, `robust` `cluster(`*varname*`)`]

`ml check`

`ml search` [[/]*eqname*[:] $\#_{lb}$ $\#_{ub}$] [[/]*eqname*[:] $\#_{lb}$ $\#_{ub}$] [...]

`ml plot` [*eqname*:]*name* [# [# [#]]]

`ml maximize`

`ml graph`

`ml display`

where *method* is { `lf` | `d0` | `d1` | `d1debug` | `d2` | `d2debug` }.

eq is the equation to be estimated, enclosed in parentheses, and optionally with a name to be given to the equation, preceded by a colon:

  ([*eqname*:] *depvarnames* = *varnames* [, `noconstant`])
  ([*eqname*:] *varnames* [, `noconstant`])
  ([*eqname*:] *depvarnames* =)
  ([*eqname*:] [, `noconstant`])

or *eq* is the name of a parameter such as sigma with a slash in front

  /*eqname*   which is equivalent to   (*eqname*:)

`fweights`, `pweights`, `aweights`, and `iweights` are allowed, see [U] **14.1.6 weight** in the *Stata User's Guide*. With all but method lf, you must write your likelihood-evaluation program a certain way if `pweights` are to be specified, and `pweights` may not be specified with method d0.

`ml` shares the features of all estimation commands; see [U] **23 Estimation and post-estimation commands** in the *Stata User's Guide*. To redisplay results, use `ml display`.

2.2 Using ml

`ml` is the Stata command to maximize user-defined likelihoods. To obtain maximum likelihood estimates,

1. You write a program—the name is of your choosing—that calculates the log-likelihood values and, optionally, its derivatives. The program you write is known as the likelihood evaluator.
2. You define a particular model to be estimated using the `ml model` statement. (In Step 1 you programmed the likelihood function in a generic way, implementing, say, probit models. At this step, you state the specific probit model to be estimated, such as diseased as a function of age and exposure.)
3. You estimate the model using `ml maximize`.

Between Steps 2 and 3 you can

2.1 Use `ml check` to verify the program you wrote has no syntax errors.

2.2 Use `ml init` to specify initial values.

2.3 Additionally or instead use `ml search` to hunt for initial values.

2.4 Additionally or instead use `ml plot` to graph the log-likelihood function and find initial values.

What you do about initial values is up to you. In most cases, `ml maximize` will succeed even if the log-likelihood function cannot be evaluated at the default initial values (which are a vector of zeros). Providing better initial values usually speeds maximization.

After Step 3, you can

3.1 Graph the log-likelihood iteration values using `ml graph`.

3.2 Redisplay results using `ml display`.

2.2.1 Example: Single equation

The probit log-likelihood function for the jth observation is

$$\ln \ell_j = \begin{cases} \ln \Phi(\theta_j) & \text{if } y_j = 1 \\ \ln \Phi(-\theta_j) & \text{if } y_j = 0 \end{cases}$$

$$\theta_j = \mathbf{x}_j \boldsymbol{\beta}$$

The following is all that is required to estimate a probit model:

```
program define myprobit
        version 6
        args lnf theta
        quietly replace `lnf' = ln(normprob(`theta')) if $ML_y1==1
        quietly replace `lnf' = ln(normprob(-`theta')) if $ML_y1==0
end
. ml model lf myprobit (foreign=mpg weight)
. ml maximize
```

The results of defining this likelihood-evaluation program and issuing these two `ml` commands are

```
. ml model lf myprobit (foreign=mpg weight)

. ml maximize
initial:       log likelihood = -51.292891
alternative:   log likelihood = -45.055272
rescale:       log likelihood = -45.055272
Iteration 0:   log likelihood = -45.055272
Iteration 1:   log likelihood = -27.904114
Iteration 2:   log likelihood =  -26.85781
Iteration 3:   log likelihood = -26.844198
Iteration 4:   log likelihood = -26.844189
Iteration 5:   log likelihood = -26.844189
                                                  Number of obs   =         74
                                                  Wald chi2(2)    =      20.75
Log likelihood = -26.844189                       Prob > chi2     =     0.0000

------------------------------------------------------------------------------
 foreign |      Coef.   Std. Err.       z     P>|z|       [95% Conf. Interval]
---------+--------------------------------------------------------------------
     mpg |  -.1039503   .0515689     -2.016   0.044      -.2050235   -.0028772
  weight |  -.0023355   .0005661     -4.126   0.000       -.003445   -.0012261
   _cons |   8.275464   2.554142      3.240   0.001       3.269438    13.28149
------------------------------------------------------------------------------
```

The `myprobit` four-line program is sufficient to estimate any probit model and to report for it either the conventional inverse negative second derivative variance estimates or the robust Huber/White/sandwich variance estimates. The robust variance estimate would be obtained simply by specifying the `robust` option on the `ml model` command.

We will explain the details of the `myprobit` program later, but our programs all follow the outline

```
program define myprog
        version 6
        ...
end
```

We are using Stata version 6.0, which is the current release of Stata at the time this book is being written. Hence we code "`version 6`" at the top of our programs. You could omit the line, but we recommend you do not. Stata is continually being developed and sometimes details of syntax change. Placing `version 6` at the top of your program tells Stata that, if anything has changed, you want the version 6 interpretation. Thus your program will continue to work with future releases of Stata.

2.2.2 Example: Two equations

The linear regression log-likelihood function for the jth observation is

$$\begin{aligned} \ln \ell_j &= \ln \phi\big((y_j - \theta_{1j})/\theta_{2j}\big) - \ln \theta_{2j} \\ \theta_{1j} &= \mathbf{x}_j \boldsymbol{\beta} \\ \theta_{2j} &= \sigma \end{aligned}$$

where $\phi()$ is the unit-normal density.

```
program define myreg
        version 6
        args lnf theta1 theta2
        quietly replace `lnf' = ln(normd(($ML_y1-`theta1')/`theta2')) - ln(`theta2')
end
```

```
. ml model lf myreg (mpg=weight displ) /sigma
. ml maximize
initial:       log likelihood =     -<inf>  (could not be evaluated)
feasible:      log likelihood = -10383.274
rescale:       log likelihood = -292.89564
rescale eq:    log likelihood = -238.45986
Iteration 0:   log likelihood = -238.45986  (not concave)
Iteration 1:   log likelihood = -225.48233  (not concave)
Iteration 2:   log likelihood = -215.66639
Iteration 3:   log likelihood = -199.18578
Iteration 4:   log likelihood = -195.29823
Iteration 5:   log likelihood = -195.23999
Iteration 6:   log likelihood =  -195.2398
Iteration 7:   log likelihood =  -195.2398
                                                  Number of obs   =         74
                                                  Wald chi2(2)    =     139.21
Log likelihood =  -195.2398                       Prob > chi2     =     0.0000

------------------------------------------------------------------------------
     mpg |      Coef.   Std. Err.       z     P>|z|       [95% Conf. Interval]
---------+--------------------------------------------------------------------
eq1      |
  weight |  -.0065671   .0011424     -5.749   0.000       -.0088061  -.0043281
   displ |   .0052808   .0096674      0.546   0.585       -.0136671   .0242286
   _cons |   40.08452   1.978738     20.258   0.000        36.20627   43.96278
---------+--------------------------------------------------------------------
sigma    |
   _cons |   3.385282   .2782684     12.166   0.000        2.839886   3.930678
------------------------------------------------------------------------------
```

Notes:

1. Most people think of linear regression as a single-equation model, $y_j = \mathbf{x}_j\boldsymbol{\beta}$, with an ancillary parameter σ also to be estimated. This is just a matter of jargon and, in `ml` jargon, the ancillary parameter σ counts as an equation.

2. We specified the "equation" for θ_2 as `/sigma` in the `ml model` statement. We could equally well have specified it as `()`:

   ```
   . ml model lf myreg (mpg=weight displ) ()
   ```

   Each θ_{ij} has a corresponding definition, $\theta_{ij} = \mathbf{x}_{ij}\boldsymbol{\beta}_i$, and these are listed one after the other on the `ml model` statement. When a definition lists no variables, it means that the corresponding $\mathbf{x}_i$ contains only a column of 1s. Specifying `/sigma` literally means `(sigma:)`. What appears in front of the colon is called the equation's name and it plays no role except to label the output. When you type `()`, `ml` uses a default name to label the equation.

3. This really is a two-equation model. We would type

   ```
   . ml model lf myreg (mpg=weight displ) (sigma: price)
   ```

   to specify that the equation for σ is $\gamma_0 + \gamma_1$`price`.

2.2.3 Example: Two equations, two dependent variables

The Weibull log-likelihood function for the jth observation, where t_j is the time of failure or censoring and $d_j = 1$ if failure and 0 if censored, is

$$\ln \ell_j = (t_j e^{\theta_{1j}})^{\exp(\theta_{2j})} + d_j\big(\theta_{2j} - \theta_{1j} + (e^{\theta_{2j}} - 1)(\ln t_j - \theta_{1j})\big)$$
$$\theta_{1j} = \mathbf{x}_j\boldsymbol{\beta}$$
$$\theta_{2j} = s$$

```
program define myweib
        version 6
        args lnf theta1 theta2
        tempvar p M R
        quietly gen double `p' = exp(`theta2')
        quietly gen double `M' = ($ML_y1*exp(-`theta1'))^`p'
        quietly gen double `R' = ln($ML_y1)-`theta1'
        quietly replace `lnf' = -`M' + $ML_y2*(`theta2'-`theta1' + (`p'-1)*`R')
end
```

```
. ml model lf myweib (studytim died = drug2 drug3 age) /s
. ml max
initial:        log likelihood =       -744
alternative:    log likelihood = -356.14276
rescale:        log likelihood = -200.80201
rescale eq:     log likelihood = -136.69232
Iteration 0:    log likelihood = -136.69232  (not concave)
Iteration 1:    log likelihood = -124.11744
Iteration 2:    log likelihood = -113.88918
Iteration 3:    log likelihood = -110.30382
Iteration 4:    log likelihood = -110.26747
Iteration 5:    log likelihood = -110.26736
Iteration 6:    log likelihood = -110.26736
                                                  Number of obs   =         48
                                                  Wald chi2(3)    =      35.25
Log likelihood = -110.26736                       Prob > chi2     =     0.0000

------------------------------------------------------------------------------
         |      Coef.   Std. Err.       z     P>|z|       [95% Conf. Interval]
---------+--------------------------------------------------------------------
eq1      |
   drug2 |   1.012966   .2903917      3.488   0.000       .4438086    1.582123
   drug3 |    1.45917   .2821195      5.172   0.000       .9062261    2.012114
     age |  -.0671728   .0205688     -3.266   0.001      -.1074868   -.0268587
   _cons |   6.060723   1.152845      5.257   0.000       3.801188    8.320259
---------+--------------------------------------------------------------------
s        |
   _cons |   .5573333   .1402154      3.975   0.000       .2825162    .8321504
------------------------------------------------------------------------------
```

Notes:

1. This model has two dependent variables, t_j and d_j. We simply specified them on the left-hand side of the equal sign in the equation definition. The order in which we placed them—`studytim died`—determined which was `$ML_y1` and which was `$ML_y2`.
2. In writing this program, we found it convenient to use temporary variables to store intermediate results. It is of great importance that those temporary variables were generated as `doubles`.

2.3 The jargon of ml

Any likelihood function may be written as

$$\begin{aligned}\ln L &= \ln L\big((\theta_{1j},\theta_{2j},\ldots,\theta_{Ej};y_{1j},y_{2j},\ldots,y_{Dj}) : j=1,2,\ldots,N\big)\\ &= \ln L\big(\boldsymbol{\theta}_1,\boldsymbol{\theta}_2,\ldots,\boldsymbol{\theta}_E;\mathbf{y}_1,\mathbf{y}_2,\ldots,\mathbf{y}_D\big)\end{aligned}$$

where j indexes the observations and $\theta_{1j}=\mathbf{x}_{1j}\boldsymbol{\beta}_1$, $\theta_{2j}=\mathbf{x}_{2j}\boldsymbol{\beta}_2$, …, $\theta_{Ej}=\mathbf{x}_{Ej}\boldsymbol{\beta}_E$.

In the notation above,

$\boldsymbol{\beta}=(\boldsymbol{\beta}_1,\boldsymbol{\beta}_2,\ldots,\boldsymbol{\beta}_E)$	are the coefficients to be estimated,
E, $E\geq 1$	is said to be the number of equations, and
D, $D\geq 0$	is said to be the number of dependent variables.

Note that $\boldsymbol{\beta}_1$, $\boldsymbol{\beta}_2$, …, $\boldsymbol{\beta}_E$ are each potentially vectors.

In the special case when

$$\begin{aligned}\ln \ell_j &= \ln \ell(\theta_{1j},\theta_{2j},\ldots,\theta_{Ej};y_{1j},y_{2j},\ldots,y_{Dj})\\ \ln L &= \sum_{j=1}^{N}\ln \ell_j\end{aligned}$$

that is, when (a) the log-likelihood contribution can be calculated separately for each observation and (b) the sum of the individual contributions equals the overall log likelihood, the likelihood function is said to meet the linear-form restrictions.

For instance, the probit likelihood function meets the linear-form restrictions: the likelihood value can be calculated for each observation in absence of knowledge of the other observations and the overall log likelihood is just the sum of the individually calculated contributions. Conditional logistic regression, on the other hand, violates the linear-form requirements because log-likelihood values can only be calculated for groups of observations.

2.4 Choosing a maximization method

`ml` provides four methods for maximizing likelihood functions:

1. Method lf is for use when (1) the likelihood function meets the linear-form restrictions and (2) you do not wish to code first or second derivatives.
2. Method d0 is for use with any likelihood function—not just those meeting the linear-form restrictions—and as with method lf, method d0 does not require that you code first or second derivatives.
3. Method d1 is for use with any likelihood function when you code first derivatives.
4. Method d2 is for use with any likelihood function when you code both first and second derivatives.

You specify the method and the name of the program you write to evaluate the likelihood function on the `ml model` statement:

`ml model` *method progname* ...

We heartily recommend method lf. It is easy to program, quick to execute, and accurate. By comparison, method d0 is slow and cumbersome, but it handles problems that method lf will not. Our recommendation is to use method lf whenever possible and resort to method d0 when forced.

The other reason to use method d0 is that you really want to use method d1 or d2 and writing a method d0 evaluator is a step toward the ultimate goal. Method d2 provides the fastest execution and the most accurate results, but it is also the most difficult to program. We do not recommend that you use method d2 unless you plan on using the resulting estimator often; otherwise, the amount of time you will invest in writing the program will not be justified.

In our experience, method d1 is seldom a happy compromise. If you can write a method d1 evaluator, in most circumstances it is not that much more work to write a method d2 evaluator.

		The four methods rated by		
	Ranking	Ease of use	Accuracy	Speed
Best	1	`lf`	`d2`	`d2`
	2	`d0`	`lf`	`lf`
	3	`d1`	`d1`	`d1`
Worst	4	`d2`	`d0`	`d0`

2.5 Specifying equations

The key to understanding `ml` is understanding `ml`'s equation–θ notation. For instance, most people would write the log likelihood for the linear regression model as

$$\ln L = \sum_{j=1}^{N} \ln \phi\big((y_j - \mathbf{x}_j\boldsymbol{\beta})/\sigma\big) - \ln \sigma$$

where $(\boldsymbol{\beta}, \sigma)$ is to be estimated and $\phi()$ is the unit-normal density. `ml` wants you to write it as

$$\begin{aligned} \ln L &= \sum_{j=1}^{N} \ln \phi\big((y_j - \theta_{1j})/\theta_{2j}\big) - \ln \theta_{2j} \\ \theta_{1j} &= \mathbf{x}_j\boldsymbol{\beta} \\ \theta_{2j} &= \sigma \end{aligned}$$

The idea here is that you will think about (and code) your likelihood-evaluation program in terms of θ_1 and θ_2. Later, when it comes time to estimate a model, you will specify linear equations for each of the thetas and `ml` will produce the corresponding estimates.

You write a program to evaluate the likelihood function:

```
program define myreg
        This program receives θ1j and θ2j as arguments and yj in $ML_y1.
        It returns ln L as output.
end
```

You then specify the particular model to be estimated by typing

```
. ml model lf myprog (equation for θ1) (equation for θ2)
```

and you finally obtain estimates by typing

```
. ml maximize
```

For instance, if you had already defined the likelihood evaluation program **myreg**, typing

```
. ml model lf myreg (mpg=weight displ) ()
. ml maximize
```

would obtain estimates of a linear regression of **mpg** on variables **weight** and **displ**. Look at the **ml model** statement. The part that reads '**(mpg=weight displ)**' specifies

1. y_j is variable **mpg**.
2. $\theta_{1j} = \beta_1 \texttt{weight}_j + \beta_2 \texttt{displ}_j + \beta_3$.

The odd looking **()** at the end of the **ml model** statement specifies

3. $\theta_{2j} = \beta_4$

ml now has everything it needs to know. It has a program **myreg** that defines $\ln L(\theta_{1j}, \theta_{2j}; y_j)$ and it has definitions for y_j, θ_{1j}, and θ_{2j}. We have not yet discussed all the details, but let us show that this will work. First, here is the real program that will evaluate the linear regression likelihood function:

```
program define myreg
        version 6
        args lnf theta1 theta2
        quietly replace `lnf'=ln(normd(($ML_y1-`theta1')/`theta2')) - ln(`theta2')
end
```

and here is the result of typing the **ml model** followed by **ml maximize** commands:

```
. ml model lf myreg (mpg=weight displ) ()
. ml maximize
initial:       log likelihood =     -<inf>  (could not be evaluated)
feasible:      log likelihood = -10383.274
rescale:       log likelihood = -292.89564
rescale eq:    log likelihood = -238.45986
Iteration 0:   log likelihood = -238.45986  (not concave)
Iteration 1:   log likelihood = -225.38874  (not concave)
Iteration 2:   log likelihood = -218.89112
Iteration 3:   log likelihood = -204.77474
Iteration 4:   log likelihood = -195.24136
Iteration 5:   log likelihood =  -195.2398
Iteration 6:   log likelihood =  -195.2398
                                                  Number of obs   =         74
                                                  Wald chi2(2)    =     139.21
Log likelihood =  -195.2398                       Prob > chi2     =     0.0000

------------------------------------------------------------------------------
     mpg |      Coef.   Std. Err.       z     P>|z|       [95% Conf. Interval]
---------+--------------------------------------------------------------------
eq1      |
  weight |  -.0065671   .0011424     -5.749   0.000      -.0088061   -.0043281
   displ |   .0052808   .0096675      0.546   0.585      -.0136671    .0242286
   _cons |   40.08452   1.978739     20.258   0.000       36.20627    43.96278
---------+--------------------------------------------------------------------
eq2      |
   _cons |   3.385282   .2782685     12.166   0.000       2.839886    3.930678
------------------------------------------------------------------------------
```

Now let's back up and understand what happened. In the output above, `eq1` stands for equation 1 and that corresponds to θ_1. `eq2` stands for equation 2 and that corresponds to θ_2. We just estimated

$$
\begin{aligned}
\ln L &= \sum_{j=1}^{N} \ln \phi\big((y_j - \theta_{1j})/\theta_{2j}\big) - \ln \theta_{2j} \\
y_j &= \texttt{mpg}_j \\
\theta_{1j} &= \beta_1 \texttt{weight}_j + \beta_2 \texttt{displ}_j + \beta_3 \\
\theta_{2j} &= \beta_4
\end{aligned}
$$

because we wrote a program to calculate

$$
\ln L = \sum_{j=1}^{N} \ln \phi\big((y_j - \theta_{1j})/\theta_{2j}\big) - \ln \theta_{2j}
$$

and we defined y_j, θ_{1j}, and θ_{2j} by typing

```
. ml model lf myreg (mpg=weight displ) ()
```

In general, any likelihood problem can be written as

$$
\begin{aligned}
\ln L &= \ln L\big((\theta_{1j}, \theta_{2j}, \dots, \theta_{Ej}; y_{1j}, y_{2j}, \dots, y_{Dj}) : \; j = 1, 2, \dots, N\big) \\
y_{1j} &= \dots \\
y_{2j} &= \dots \\
&\vdots \\
y_{Dj} &= \dots \\
\theta_{1j} &= \dots \\
\theta_{2j} &= \dots \\
&\vdots \\
\theta_{Ej} &= \dots
\end{aligned}
$$

The program you write fills in the definition of $\ln L$. The `ml model` statement you issue fills in the dots.

Equations are specified on the `ml model` statement

`. ml model ...` (*equation for* θ_1) (*equation for* θ_2) `...`

The syntax for specifying an equation is

([*equation_name*:] [*varname(s)* =] *varname(s)* [, <u>`noconst`</u>`ant`])

equation_name is optional: if you do not specify it, the first equation will be named `eq1`, the second `eq2`, etc. Specifying an equation name only makes the output look prettier. Had we, in our linear regression example, specified the equation for θ_{2j} as `(sigma:)` rather than `()`, the output would have shown the word `sigma` where it now has `eq2`.

The first set of *varname(s)* in front of the equal sign specifies the y variables. The second set of *varname(s)* specifies the variables on which the linear equation for θ is to depend. Finally, whether we type `noconstant` determines whether the linear equation includes a constant.

Forget the linear regression example for a minute. Without the context of a particular likelihood function, here are some equation specifications and their translation:

```
. ml model ... (foreign=mpg weight)
```

$$y_{1j} = \texttt{foreign}_j$$
$$\theta_{1j} = \beta_1\texttt{mpg}_j + \beta_2\texttt{weight}_j + \beta_3$$

```
. ml model ... (mpg=weight displ) ()
```

$$y_{1j} = \texttt{mpg}_j$$
$$\theta_{1j} = \beta_1\texttt{weight}_j + \beta_2\texttt{displ}_j + \beta_3$$
$$\theta_{2j} = \beta_4$$

```
. ml model ... (mpg=weight displ) (price)
```

$$y_{1j} = \texttt{mpg}_j$$
$$\theta_{1j} = \beta_1\texttt{weight}_j + \beta_2\texttt{displ}_j + \beta_3$$
$$\theta_{2j} = \beta_4\texttt{price}_j + \beta_5$$

```
. ml model ... (mpg=weight displ, nocons) (price)
```

$$y_{1j} = \texttt{mpg}_j$$
$$\theta_{1j} = \beta_1\texttt{weight}_j + \beta_2\texttt{displ}_j$$
$$\theta_{2j} = \beta_3\texttt{price}_j + \beta_4$$

```
. ml model ... (mpg=weight displ) (price, nocons)
```

$$y_{1j} = \texttt{mpg}_j$$
$$\theta_{1j} = \beta_1\texttt{weight}_j + \beta_2\texttt{displ}_j + \beta_3$$
$$\theta_{2j} = \beta_4\texttt{price}_j$$

```
. ml model ... (mpg=weight displ) (price=weight) ()
```

$$
\begin{aligned}
y_{1j} &= \texttt{mpg}_j \\
y_{2j} &= \texttt{price}_j \\
\theta_{1j} &= \beta_1 \texttt{weight}_j + \beta_2 \texttt{displ}_j + \beta_3 \\
\theta_{2j} &= \beta_4 \texttt{weight}_j + \beta_5 \\
\theta_{3j} &= \beta_6
\end{aligned}
$$

```
. ml model ... (mpg price=weight displ) (weight) ()
```

$$
\begin{aligned}
y_{1j} &= \texttt{mpg}_j \\
y_{2j} &= \texttt{price}_j \\
\theta_{1j} &= \beta_1 \texttt{weight}_j + \beta_2 \texttt{displ}_j + \beta_3 \\
\theta_{2j} &= \beta_4 \texttt{weight}_j + \beta_5 \\
\theta_{3j} &= \beta_6
\end{aligned}
$$

That is how equation specification works. Note especially the last two examples:

```
. ml model ... (mpg=weight displ) (price=weight) ()
. ml model ... (mpg price=weight displ) (weight) ()
```

Both resulted in the same definitions of y_{1j}, y_{2j}, θ_{1j}, θ_{2j}, and θ_{3j}. When you specify the "dependent" (y) variables, it really does not matter on which equation you specify them. The first y variable becomes y_1, the second y_2, and so on.

In our examples above we never bothered to specify the equation name. Let us go back to our linear regression example and show you the result of specifying it:

```
. ml model lf myreg (mpg=weight displ) (sigma:)
. ml maximize
initial:       log likelihood =     -<inf>  (could not be evaluated)
feasible:      log likelihood = -10383.274
  (output omitted)
Iteration 6:   log likelihood =  -195.2398
                                                  Number of obs   =         74
                                                  Wald chi2(2)    =     139.21
Log likelihood =  -195.2398                       Prob > chi2     =     0.0000

------------------------------------------------------------------------------
     mpg |      Coef.   Std. Err.       z     P>|z|       [95% Conf. Interval]
---------+--------------------------------------------------------------------
eq1      |
  weight |  -.0065671   .0011424     -5.749   0.000       -.0088061   -.0043281
   displ |   .0052808   .0096675      0.546   0.585       -.0136671    .0242286
   _cons |   40.08452   1.978739     20.258   0.000        36.20627    43.96278
---------+--------------------------------------------------------------------
sigma    |
   _cons |   3.385282   .2782685     12.166   0.000        2.839886    3.930678
------------------------------------------------------------------------------
```

As we said earlier, the result was to change the eq2 at the bottom of the output to sigma. That looks better, but that is all. Had we specified an equation name for the first equation, say by typing

```
. ml model lf myreg (mpg: mpg=weight displ) (sigma:)
```

then, rather than the output being labeled eq1 for the first equation, it would have been labeled mpg.

Actually, specifying an equation name does more than just change how the output is labeled. If, later in an expression—well after the model has been estimated—we wished to refer to the coefficient on weight, we would properly refer to [*equation_name*]_b[weight], so we would type [eq1]_b[weight] in one case and [mpg]_b[weight] in another. In fact, we would probably just type _b[weight] because, when we do not specify the equation, Stata's expression parser assumes we are referring to the first equation, but the point still remains when we wish to refer to a coefficient in the second, third, ..., equation.

There are other ways that ml model will let us specify equations. For instance, instead of typing (*equation_name*:), which is one way of specifying a constant and a name, we can type /*equation_name*. That actually looks nice and we tend to do that. For instance, in our linear regression example, we would probably specify the model as

```
. ml model lf myreg (mpg=weight displ) /sigma
```

rather than

```
. ml model lf myreg (mpg=weight displ) (sigma:)
```

or

```
. ml model lf myreg (mpg=weight displ) ()
```

Which we do really makes no substantive difference, however.

Now that you understand equations, let us show the real power of the notation. Without explaining the likelihood-evaluation program itself, we have shown you that

```
program define myreg
        version 6
        args lnf theta1 theta2
        quietly replace `lnf' = ln(normd(($ML_y1-`theta1')/`theta2')) - ln(`theta2')
end
. ml model lf myreg (mpg=weight displ) /sigma
. ml maximize
```

will produce the maximum likelihood estimates of

$$\texttt{mpg}_j = \beta_1\texttt{weight}_j + \beta_2\texttt{displ}_j + \beta_3 + \epsilon_j, \quad \epsilon_j \sim N(0, \sigma^2)$$

Without changing our likelihood-evaluation program myreg at all, we can just as easily estimate a model in which σ is not constant but is allowed to vary. For instance, we could assume $\sigma_j = \gamma_1\texttt{weight}_j + \gamma_2$ by typing

```
. ml model lf myreg (mpg=weight displ) (sigma: weight)
. ml maximize
initial:       log likelihood =     -<inf>  (could not be evaluated)
feasible:      log likelihood = -10383.274
rescale:       log likelihood = -292.89564
rescale eq:    log likelihood = -238.45986
Iteration 0:   log likelihood = -238.45986  (not concave)
Iteration 1:   log likelihood = -223.69833  (not concave)
Iteration 2:   log likelihood = -210.11509
Iteration 3:   log likelihood = -195.58018
```

```
Iteration 4:   log likelihood = -190.59899
Iteration 5:   log likelihood = -190.20701
Iteration 6:   log likelihood = -190.19848
Iteration 7:   log likelihood = -190.19848
                                                  Number of obs   =         74
                                                  Wald chi2(2)    =     140.31
Log likelihood = -190.19848                       Prob > chi2     =     0.0000

------------------------------------------------------------------------------
     mpg |      Coef.   Std. Err.       z     P>|z|       [95% Conf. Interval]
---------+--------------------------------------------------------------------
eq1      |
  weight |  -.0060495   .0011241     -5.382   0.000      -.0082526   -.0038463
   displ |   .0052541    .008891      0.591   0.555      -.0121718    .0226801
   _cons |   38.47725   2.095853     18.359   0.000       34.36946    42.58505
---------+--------------------------------------------------------------------
sigma    |
  weight |  -.0009508   .0002892     -3.288   0.001      -.0015176    -.000384
   _cons |   6.122385   1.026742      5.963   0.000       4.110008    8.134763
------------------------------------------------------------------------------
```

2.6 Specifying a subsample

If you wish to estimate your model on a subsample of the data, for instance, the data for which `trunk>8`, you specify the restriction on the `ml model` statement. Continuing to use our linear regression example, typing

```
. ml model lf myreg (mpg=weight displ) /sigma if trunk>8
. ml maximize
```

would estimate the model on a subset of the observations. You specify either or both `if` *exp* and `in` *range*.

2.7 Specifying weights

If you wish to perform weighted estimation, you specify the weights on the `ml model` statement. For instance, using our linear regression example but with another dataset, we might type

```
. ml model lf myreg (lninc = ed exp) /sigma [pweight=pop]
. ml maximize
```

You may specify `fweights`, `pweights`, `aweights`, or `iweights`. There is one issue concerning `pweights`: If you use method d0, you may not specify `pweights` and if you use method d1 or d2, you must write your program in a special way. This will be explained in **6.3 Robust variance estimates** (robust variance estimates are how one handles `pweights`). There is nothing special you need do, however, when using method lf.

2.8 Obtaining robust estimates of variance

Robust estimates of variance were explained in **1.3.4 Robust variance estimates**. If you wish to obtain the robust estimate of variance, you specify the `robust` and/or `cluster()` option on the `ml model` statement:

```
. ml model lf myreg (mpg=weight displ) /sigma, robust
. ml maximize
```

There is an issue if you are not using method lf and this is the same issue as concerning `pweights`. If you use method d0, you may not specify `robust` or `cluster()` and if you use method d1 or d2, you must write your program in a special way.

2.9 Maximizing your own functions

1. Enter your likelihood-evaluation program:

   ```
   program define myprog
           Program receives θ1j, θ2j, ..., as arguments and y1j in $ML_y1, y2j in $ML_y2, ....
           It returns ln L as output.
   end
   ```

   See **3 Method lf**, **5 Method d0**, **6 Method d1**, or **7 Method d2** according to which method you use.

2. Specify the method (lf, d0, d1, or d2), the name of your program (*myprog*), and the model to be estimated; type

   ```
   . ml model method myprog (equation for θ1) (equation for θ2) ...
   ```

   It is at this point that you specify sample inclusion (`if` *exp* and `in` *range*), weights, and options such as `robust` and `cluster()` for the robust Huber/White/sandwich variance estimate.

3. Verify the program you have written is free of syntax errors and other common problems; type

   ```
   . ml check
   ```

   See **8.2 ml check** if you wish, but the output produced by `ml check` is obvious.

4. Optionally specify starting values using `ml init`; see **9.4 ml init**. We do not recommend this unless you have an analytic approximation that gives good starting values. Otherwise, use `ml search`; see Step 6 below.

5. Optionally look for starting values using `ml plot`; see **9.3 ml plot**. This is more cute than useful because `ml search` (see step 6) works better.

6. Optionally—and we recommend this—tell `ml` to search for starting values; type

   ```
   . ml search
   ```

   See **9.2 ml search**.

7. Obtain maximum likelihood estimates; type

   ```
   . ml maximize
   ```

   See **10 Interactive maximization**.

8. If you wish to graph the convergence, type

   ```
   . ml graph
   ```

   See **11.2 Graphing convergence**.

9. If, at a later time, you wish to redisplay the last estimates, type

   ```
   . ml display
   ```

   See **11.3 Redisplaying output**.

10. If you wish to add a new command to Stata to perform this kind of estimation (this is a bit of work), see **13 Writing ado-files to maximize likelihoods**.

3 Method lf

Contents

3.1 Basic syntax

Method lf is for use when (1) the likelihood function meets the linear-form restrictions and (2) you do not wish to code first and second derivatives. We urge that you use lf whenever possible. Method lf is fast and accurate; only method d2 will beat it on either score.

The likelihood function is assumed to be of the form

$$
\begin{aligned}
\ln \ell_j &= \ln \ell_j(\theta_{1j}, \theta_{2j}, \ldots, \theta_{Ej}; y_{1j}, y_{2j}, \ldots, y_{Dj}) \\
\ln L &= \sum_{j=1}^{N} \ln \ell_j \\
\theta_{1j} &= \mathbf{x}_{1j}\boldsymbol{\beta}_1 \\
\theta_{2j} &= \mathbf{x}_{2j}\boldsymbol{\beta}_2 \\
&\vdots \\
\theta_{Ej} &= \mathbf{x}_{Ej}\boldsymbol{\beta}_E
\end{aligned}
$$

where j indexes observations.

The method lf evaluator is to calculate $\ln \ell_j$, the log likelihood observation-by-observation. The outline for the program is

```
program define myprog
        version 6
        args lnf theta1 theta2 ...
        /* if you need to create any intermediate results: */
        tempvar tmp1 tmp2 ...
        quietly gen double `tmp1' = ...
        ...
        quietly replace `lnf' = ...
end
```

You specify that you are using method lf with program *myprog* on the `ml model` statement,

```
. ml model lf myprog ...
```

which you issue after you have defined the likelihood-evaluation program.

Notes:

1. Access the dependent variables (if any) by typing `$ML_y1`, `$ML_y2`, ... Use `$ML_y1`, `$ML_y2`, ..., just as you would any existing variable in the data.
2. Final results are to be saved in `` `lnf' ``. `` `lnf' `` is just a double-precision variable `ml` created for you. When `ml` calls your program, `` `lnf' `` contains missing values.
3. It is of great importance that, if you create any temporary variables for intermediate results, you create them as `doubles`. Type

   ```
   . gen double name = ...
   ```

   Do not omit the word `double`.
4. You must not change the contents of variables `` `theta1' ``, `` `theta2' ``, ... Never code

   ```
   . replace `theta1' = ...
   ```

Using method lf is simply a matter of writing down the likelihood function for an observation, changing the mathematical notation to `ml`'s equation–θ notation, and coding it.

3.1.1 The linear-form restrictions

Method lf requires that the physical observations in the dataset correspond to independent pieces of the likelihood function:

$$\ln L = \ln \ell_1 + \ln \ell_2 + \cdots + \ln \ell_N$$

This restriction is commonly (but not always) met in statistical analyses. It corresponds to

$$\Pr(\text{dataset}) = \Pr(\text{obs. } 1) \times \Pr(\text{obs. } 2) \times \cdots \times \Pr(\text{obs. } N)$$

with discrete data. Anytime you have a likelihood function and can write it as

"The log likelihood for the jth observation is $\ln \ell_j = \ldots$"

and that formula uses data for observation j only, and it is obvious that

"The overall log-likelihood function is $\ln L = \sum_{j=1}^{N} \ln \ell_j$."

you have a likelihood meeting this restriction. In other words, the observations are independent.

The following are examples of likelihoods that *do not* meet this criterion: the conditional (fixed-effects) logit model, Cox regression, and likelihoods for panel data (i.e., cross-sectional time-series models). Basically, any likelihood that explicitly models independent groups rather than independent observations does not meet the lf criterion.

Let us consider examples that do meet the lf criterion.

3.1.2 Example: A single-equation model

The probit log-likelihood function for the jth observation is

$$\ln \ell_j = \begin{cases} \ln \Phi(\theta_j) & \text{if } y_j = 1 \\ \ln \Phi(-\theta_j) & \text{if } y_j = 0 \end{cases}$$

$$\theta_j = \mathbf{x}_j \boldsymbol{\beta}$$

The following is the method lf evaluator for this likelihood function:

```
program define myprobit
        version 6
        args lnf theta
        quietly replace `lnf' = ln(normprob(`theta')) if $ML_y1==1
        quietly replace `lnf' = ln(normprob(-`theta')) if $ML_y1==0
end
```

The evaluation program receives two arguments because this is a one-equation model. The first argument is the log likelihood which we are to fill in and the second is θ. y, the dependent variable, is `$ML_y1`.

If we wanted to estimate a model of `foreign` on `mpg` and `weight`, we would type

```
. ml model lf myprobit (foreign=mpg weight)
. ml maximize
```

The '`foreign=`' part specifies that y is `foreign`. The '`mpg weight`' part specifies that

$$\theta_j = \beta_1 \texttt{mpg}_j + \beta_2 \texttt{weight}_j + \beta_3$$

Understand that our program never has to deal with the fact that $\theta_j = \beta_1 \texttt{mpg}_j + \beta_2 \texttt{weight}_j + \beta_3$ nor, indeed, is it ever even aware of it. Let's pretend that at some point in the maximization process, $\beta_1 = -.1$, $\beta_2 = -.002$, and $\beta_3 = 8$. We never have to know any of that because all lf will ask our program is questions like

What is the log likelihood if outcome is 1 and $\theta_j = .5$?

and all we have to do is reply "$-.36894642$", which is $\ln \Phi(.5)$, or

What is the log likelihood if outcome is 0 and $\theta_j = .1$?

to which we reply, "$-.77615459$", which is $\ln \Phi(-.1)$. In fact, lf asks both these questions at once, along with many others, when it asks us to fill in the dataset variable `` `lnf' `` with the log likelihood for each of the observations based on the values of the variables in `$ML_y1` and `` `theta' ``. The way this works, we never need to know that the particular model being estimated contains `mpg` and `weight`, or whether the model has an intercept.

Here is the result of using `myprobit` to estimate a model of `foreign` on `mpg` and `weight`

```
. ml model lf myprobit (foreign=mpg weight)

. ml maximize

initial:       log likelihood = -51.292891
alternative:   log likelihood = -45.055272
rescale:       log likelihood = -45.055272
Iteration 0:   log likelihood = -45.055272
Iteration 1:   log likelihood = -27.904114
Iteration 2:   log likelihood = -26.858137
Iteration 3:   log likelihood = -26.844198
Iteration 4:   log likelihood = -26.844189
Iteration 5:   log likelihood = -26.844189
```

```
                                                  Number of obs   =         74
                                                  Wald chi2(2)    =      20.75
Log likelihood = -26.844189                       Prob > chi2     =     0.0000

------------------------------------------------------------------------------
 foreign |      Coef.   Std. Err.       z     P>|z|       [95% Conf. Interval]
---------+--------------------------------------------------------------------
     mpg |  -.1039503   .0515689     -2.016   0.044      -.2050235   -.0028772
  weight |  -.0023355   .0005661     -4.126   0.000       -.003445   -.0012261
   _cons |   8.275464   2.554142      3.240   0.001       3.269437    13.28149
------------------------------------------------------------------------------
```

3.1.3 Example: A two-equation model

The linear-regression log-likelihood function for the jth observation is

$$\ln \ell_j = \ln \phi\big((y_j - \theta_{1j})/\theta_{2j}\big) - \ln \theta_{2j}$$
$$\theta_{1j} = \mathbf{x}_j \boldsymbol{\beta}$$
$$\theta_{2j} = \sigma$$

The corresponding method lf evaluator for this likelihood function is

```
program define myreg
        version 6
        args lnf theta1 theta2
        quietly replace `lnf´ = ln(normd(($ML_y1-`theta1´)/`theta2´)) - ln(`theta2´)
end
```

This function receives three arguments because there are two equations, one for θ_1 and a second for θ_2 even though, as we intend to use this function, the equation for θ_2 will be specified as a constant. For example, if we wanted to estimate a model of mpg on weight and displ with constant σ, we would type

```
. ml model lf myreg (mpg=weight displ) /sigma
. ml maximize
```

Typing "mpg=" states that $y_j = \texttt{mpg}_j$. Typing "weight displ" specifies $\theta_{1j} = \beta_1\texttt{weight}_j + \beta_2\texttt{displ}_j + \beta_3$. Typing "/sigma" for the second equation specifies $\theta_{2j} = \beta_4$.

This routine can just as easily estimate heteroscedastic models because we could specify an equation for the σ, such as $\theta_{2j} = \beta_4\texttt{weight}_j + \beta_5$. We could do that by typing

```
. ml model lf myreg (mpg=weight displ) (sigma: weight)
. ml maximize
initial:       log likelihood =     -<inf>  (could not be evaluated)
feasible:      log likelihood = -10383.274
rescale:       log likelihood = -292.89564
rescale eq:    log likelihood = -238.45986
Iteration 0:   log likelihood = -238.45986  (not concave)
Iteration 1:   log likelihood = -223.70054  (not concave)
Iteration 2:   log likelihood = -208.95298
Iteration 3:   log likelihood = -195.25053
Iteration 4:   log likelihood =  -190.4807
Iteration 5:   log likelihood = -190.19891
Iteration 6:   log likelihood = -190.19848
Iteration 7:   log likelihood = -190.19848
```

```
                                                  Number of obs   =         74
                                                  Wald chi2(2)    =     140.31
Log likelihood = -190.19848                       Prob > chi2     =     0.0000

------------------------------------------------------------------------------
     mpg |      Coef.   Std. Err.       z     P>|z|       [95% Conf. Interval]
---------+--------------------------------------------------------------------
eq1      |
  weight |  -.0060495   .0011241     -5.382   0.000      -.0082526   -.0038463
   displ |   .0052541    .008891      0.591   0.555      -.0121718    .0226801
   _cons |   38.47725   2.095853     18.359   0.000       34.36946    42.58505
---------+--------------------------------------------------------------------
sigma    |
  weight |  -.0009508   .0002892     -3.288   0.001      -.0015176    -.000384
   _cons |   6.122385   1.026743      5.963   0.000       4.110006    8.134763
------------------------------------------------------------------------------
```

Had we wanted to constrain $\beta_5 = 0$ so that $\theta_{2j} = \beta_4\texttt{weight}_j$, we would have typed

```
. ml model lf myreg (mpg=weight displ) (sigma: weight, nocons)
. ml maximize
```

Prefixing the second equation's definition with '`sigma:`' just makes the output prettier.

❑ Technical Note

The way we have parameterized this model, the log likelihood is defined only for $\theta_{2j} = \sigma > 0$. `ml` is designed to deal with restricted ranges—there is nothing special you need do—but you should be impressed and you should realize that you are asking a lot of `ml`. There could be convergence problems and, even if there are not, `ml` is going to perform more work, so the maximization process may not occur as quickly as it otherwise could.

It would be better if the model were parameterized so that $(-\infty, \infty)$ is a valid range for each of the thetas. A better way to write the linear regression likelihood function would be

$$\begin{aligned}
\ln \ell_j &= \ln \phi\big((y_j - \theta_{1j})/\exp(\theta_{2j})\big) - \theta_{2j} \\
\theta_{1j} &= \mathbf{x}_j\boldsymbol{\beta} \\
\theta_{2j} &= \ln \sigma
\end{aligned}$$

and that parameterization would lead to the evaluation program

```
program define myreg
        version 6
        args lnf theta1 theta2
        quietly replace `lnf' = ln(normd(($ML_y1-`theta1')/exp(`theta2')) - `theta2'
end
```

❑

3.1.4 Example: A two-equation, two dependent variable model

The Weibull log-likelihood function for the jth observation, where t_j is the time of failure or censoring and $d_j = 1$ if failure and 0 if censored, is

$$\begin{aligned}
\ln \ell_j &= (t_j e^{\theta_{1j}})^{\exp(\theta_{2j})} + d_j\big(\theta_{2j} - \theta_{1j} + (e^{\theta_{2j}} - 1)(\ln t_j - \theta_{1j})\big) \\
\theta_{1j} &= \mathbf{x}_j\boldsymbol{\beta} \\
\theta_{2j} &= s
\end{aligned}$$

The corresponding method lf evaluator for this likelihood function is

```
program define myweib
        version 6
        args lnf theta1 theta2

        tempvar p M R
        quietly gen double `p' = exp(`theta2')
        quietly gen double `M' = ($ML_y1*exp(-`theta1'))^`p'
        quietly gen double `R' = ln($ML_y1)-`theta1'

        quietly replace `lnf' = -`M' + $ML_y2*(`theta2'-`theta1' + (`p'-1)*`R')
end
```

This program has two dependent variables, t_j and d_j. In the code above, we just assumed that t_j would be the first variable specified (and so coded `$ML_y1` for t_j) and d_j would be the second (and so coded `$ML_y2` for d_j). We will arrange that ordering by the order in which we specify the dependent variables on the `ml model` statement:

```
. ml model lf myweib (studytim died = drug2 drug3 age) /s

. ml maximize

initial:       log likelihood =       -744
alternative:   log likelihood = -356.14276
rescale:       log likelihood = -200.80201
rescale eq:    log likelihood = -136.69232
Iteration 0:   log likelihood = -136.69232  (not concave)
Iteration 1:   log likelihood = -124.11744
Iteration 2:   log likelihood = -113.88918
Iteration 3:   log likelihood = -110.30382
Iteration 4:   log likelihood = -110.26747
Iteration 5:   log likelihood = -110.26736
Iteration 6:   log likelihood = -110.26736

                                                  Number of obs   =         48
                                                  Wald chi2(3)    =      35.25
Log likelihood = -110.26736                       Prob > chi2     =     0.0000

------------------------------------------------------------------------------
         |      Coef.   Std. Err.       z     P>|z|       [95% Conf. Interval]
---------+--------------------------------------------------------------------
eq1      |
   drug2 |   1.012966   .2903917      3.488   0.000       .4438086    1.582123
   drug3 |    1.45917   .2821195      5.172   0.000       .9062261    2.012114
     age |  -.0671728   .0205688     -3.266   0.001      -.1074868   -.0268587
   _cons |   6.060723   1.152845      5.257   0.000       3.801188    8.320259
---------+--------------------------------------------------------------------
s        |
   _cons |   .5573333   .1402154      3.975   0.000       .2825162    .8321504
------------------------------------------------------------------------------
```

3.2 The importance of generating temporary variables as doubles

In the example above we used three temporary variables in the `myweib` program:

```
quietly gen double `p' = exp(`theta2')
quietly gen double `M' = ($ML_y1*exp(-`theta1'))^`p'
quietly gen double `R' = ln($ML_y1)-`theta1'
```

We did not have to approach this problem in this way, but making the temporary variables made writing the final line easier and more readable and so reduced the chances we made an error:

```
quietly replace `lnf' = -`M' + $ML_y2*(`theta2'-`theta1' + (`p'-1)*`R')
```

Note especially that when we created these temporary variables, we created them as doubles. This is of great importance.

As the program is coded, it works. Just to show what can happen if you do not declare temporary variables to be doubles, we went back and changed the program so that the three lines read

```
quietly gen `p' = exp(`theta2')
quietly gen `M' = ($ML_y1*exp(-`theta1'))^`p'
quietly gen `R' = ln($ML_y1)-`theta1'
```

Here is what happened when we used it:

```
. ml model lf myweib (studytim died = drug2 drug3 age) /s
. ml max
initial:       log likelihood =       -744
alternative:   log likelihood = -356.14277
rescale:       log likelihood = -200.80201
rescale eq:    log likelihood = -188.26758
Iteration 0:   log likelihood = -188.26758
Iteration 1:   log likelihood = -183.36968  (not concave)
Iteration 2:   log likelihood = -183.36817  (not concave)
Iteration 3:   log likelihood = -183.22907  (not concave)
(output omitted, every line of which said "not concave")
Iteration 29:  log likelihood = -117.82058  (not concave)
Iteration 30:  log likelihood =   -117.772
Iteration 31:  log likelihood = -113.85372  (not concave)
(output omitted, every line of which said "not concave")
Iteration 51:  log likelihood = -110.37055  (not concave)
Iteration 52:  log likelihood = -110.37049
Iteration 53:  log likelihood = -110.35906  (not concave)
(output omitted, every line of which said "not concave")
Iteration 57:  log likelihood = -110.30345  (not concave)
Iteration 58:  log likelihood =  -110.3023
Iteration 59:  log likelihood = -110.30072  (not concave)
Iteration 60:  log likelihood = -110.30029  (not concave)
Iteration 61:  log likelihood = -110.30026  (not concave)
Iteration 62:  log likelihood = -110.30015  (not concave)
Iteration 63:  log likelihood = -110.30015  (not concave)
                                                  Number of obs   =         48
                                                  Wald chi2(3)    =      39.61
Log likelihood = -110.30015                       Prob > chi2     =     0.0000

------------------------------------------------------------------------------
         |      Coef.   Std. Err.       z     P>|z|       [95% Conf. Interval]
---------+--------------------------------------------------------------------
eq1      |
   drug2 |   1.036334   .2857022      3.627   0.000       .4763678      1.5963
   drug3 |   1.478277   .2793388      5.292   0.000        .930783    2.025771
     age |  -.0682136   .0210615     -3.239   0.001      -.1094934   -.0269338
   _cons |   6.110211   1.186946      5.148   0.000       3.783841    8.436582
---------+--------------------------------------------------------------------
s        |
   _cons |   .5211394          .          .       .              .           .
------------------------------------------------------------------------------
```

Note that, in the last iteration, the function was still "not concave". As we warned in Chapter 1, when the iterations end in that way, you are in trouble. It is difficult to miss because, when the last iteration is still "not concave", one or more of the standard errors will be missing.

If you forget to specify double, one of five things will happen.

1. The maximization will require lots of "not concave" iterations and, importantly, the last iteration will be "not concave". (Intermediate "not concave" iterations should not concern you.)

2. `ml maximize` will get to iteration 1 or 2 and just sit. (In that case, press the *Break* key.)
3. `ml maximize` will iterate forever, never declaring convergence.
4. There is a small chance that `ml maximize` (and `ml check`) will report that it cannot calculate the derivatives. (In most cases when you see this error, however, it is due to a different problem: The likelihood evaluates to the same constant regardless of the value of the parameter vector due to some silly error in the evaluation program; often that constant is a missing value.)
5. There is a small chance that no mechanical problem will arise, but in that case the reported standard errors are accurate to only a few digits.

If you include intermediate calculations, review the `generate` statements carefully to ensure that you have coded the word `double` following the `generate`.

3.3 Problems you can safely ignore

Method lf handles a variety of issues for you:

1. It is not your responsibility to restrict your calculations to the estimation subsample, although your program might run a little faster if you did. The variable `$ML_samp` contains 0 or 1, 1 meaning the observation is to be used and 0 meaning it is to be ignored. It is only necessary for you to fill in `` `lnf' `` in observations for which `$ML_samp==1`. Thus, in our first example, the probit likelihood-evaluation program could read

```
program define myprobit
        version 6
        args lnf theta
        quietly replace `lnf' = ln(normprob(`theta')) if $ML_y1==1 & $ML_samp==1
        quietly replace `lnf' = ln(normprob(-`theta')) if $ML_y1==0 & $ML_samp==1
end
```

   This program would execute a little more quickly in some cases, but that is all. Even if you do not do this, results reported by `ml` method lf will be correct.

2. Whether you specify `ml model`'s `nopreserve` option (see **A. Syntax diagrams**) makes no difference. This is a technical issue; do not bother to see the appendix over this. As we just said, restricting calculations to the estimation subsample is handled by `ml` method lf itself.
3. You may perform weighted estimation and there is nothing special you need do in your program. When you read about `$ML_w` in other parts of this book, ignore it if you are using `ml` method lf. Weights are handled by `ml` method lf itself.
4. You may specify the `robust` and `cluster()` options on the `ml model` statement and there is nothing special you need do in your program to handle this. (The other methods do require that your program be coded a certain way; `ml` method lf handles those issues itself.)

3.4 Nonlinear specifications

The term "linear-form restrictions" suggests the parameterization of the likelihood must be linear. That is not true. The linear-form restrictions require only that the likelihood function be calculable observation-by-observation and that the individual equations each be linear.

For instance, suppose you wished to estimate

$$y_j = \beta_1 x_{1j} + \beta_2 x_{2j} + \beta_3 x_{3j}^{\beta_4} + \beta_5 + \epsilon_j$$

where ϵ_j is distributed $N(0, \sigma^2)$. Note that the term involving x_3 is nonlinear. Here is how you could write this model:

$$
\begin{aligned}
\ln \ell_j &= \ln \phi\big((y_j - \theta_{1j} - \theta_{2j} x_{3j}^{\theta_{3j}})/\theta_{4j}\big) - \ln \theta_{4j} \\
\theta_{1j} &= \beta_1 x_{1j} + \beta_2 x_{2j} + \beta_5 \\
\theta_{2j} &= \beta_3 \\
\theta_{3j} &= \beta_4 \\
\theta_{4j} &= \sigma
\end{aligned}
$$

In `ml` jargon this is a four-equation model even though, conceptually, there is only one equation. Nonlinear terms simply add equations. The corresponding method lf evaluator for this likelihood function would be

```
program define nonlin
        version 6
        args lnf theta1 theta2 theta3 theta4
        quietly replace `lnf' = /*
        */ ln(normd($ML_y1-`theta1'-`theta2'*X3^`theta3')/`theta4' - ln(`theta4')
end
```

Note that the variable name corresponding to x_{3j} is hardcoded in the evaluation program. We used the name `X3` but you would substitute the actual variable name. For instance, if you wished to estimate $\texttt{bp} = \beta_1\texttt{age} + \beta_2\texttt{sex} + \beta_3\texttt{bp0}^{\beta_4} + \epsilon$, you would change the program to read

```
program define nonlin
        version 6
        args lnf theta1 theta2 theta3 theta4
        quietly replace `lnf' = /*
        */ ln(normd($ML_y1-`theta1'-`theta2'*bp0^`theta3')/`theta4' - ln(`theta4')
end
```

and then estimate the model by typing

```
. ml model lf nonlin (bp=age sex) /beta3 /beta4 /sigma
. ml maximize
```

Thus, the liner-form restriction is not really a linear-restriction because any nonlinear parameter can be split off into an equation that contains a constant.

3.5 The advantages of lf in terms of execution speed

The lf method provides a considerable speed advantage over the other numerical methods and, in fact, comes close to the performance of method d2. Let us show you why.

In addition to the log-likelihood function itself, the `ml` optimizer needs the first and second derivatives:

$$
\mathbf{g} = \left(\frac{\partial \ln L}{\partial \beta_1} \quad \frac{\partial \ln L}{\partial \beta_2} \quad \cdots \quad \frac{\partial \ln L}{\partial \beta_k} \right)
$$

$$
\mathbf{H} = \begin{pmatrix}
\frac{\partial^2 \ln L}{\partial \beta_1^2} & \frac{\partial^2 \ln L}{\partial \beta_1 \partial \beta_2} & \cdots & \frac{\partial^2 \ln L}{\partial \beta_1 \partial \beta_k} \\
\frac{\partial^2 \ln L}{\partial \beta_2 \partial \beta_1} & \frac{\partial^2 \ln L}{\partial \beta_2^2} & \cdots & \frac{\partial^2 \ln L}{\partial \beta_2 \partial \beta_k} \\
\vdots & \vdots & \ddots & \vdots \\
\frac{\partial^2 \ln L}{\partial \beta_k \partial \beta_1} & \frac{\partial^2 \ln L}{\partial \beta_k \partial \beta_2} & \cdots & \frac{\partial^2 \ln L}{\partial \beta_k^2}
\end{pmatrix}
$$

Note that we write $\mathbf{g}$ as a row-vector because that is what `ml` prefers.

If we do not have analytic formulas for $\mathbf{g}$ and $\mathbf{H}$, then `ml` must calculate each of these $k+k(k+1)/2$ derivatives numerically. For instance, `ml` could calculate the first element of $\mathbf{g}$ by

$$\frac{\ln L(b_1 + h/2, b_2, \ldots, b_k; \mathbf{X}) - \ln L(b_1 - h/2, b_2, \ldots, b_k; \mathbf{X})}{h}$$

This involves evaluating the log-likelihood function $\ln L()$ twice. The second element of $\mathbf{g}$ could be obtained by

$$\frac{\ln L(b_1, b_2 + h/2, \ldots, b_k; \mathbf{X}) - \ln L(b_1, b_2 - h/2, \ldots, b_k; \mathbf{X})}{h}$$

which are two more evaluations, and so on. Using centered derivatives, we would need to evaluate the log-likelihood function $2k$ times assuming we know the optimum value of h. If we have to search for an optimum h in each calculation, we will need even more evaluations of $\ln L()$.

Next we will have to evaluate the $k(k+1)/2$ second derivatives and, as we showed in **1.4.4 Numerical second derivatives**, that takes a minimum of 3 function evaluations each, although one of the evaluations is $\ln L(b_1, b_2, \ldots, b_k)$ and is in common to all the terms. Furthermore, $2k$ of the required function evaluations are in common with the function evaluations for the first derivative. Thus, we need "only" $k(k-1)$ additional function evaluations. The grand total for first and second derivatives is thus $2k + k(k-1)$.

In the case of lf, however, it is known that $\ln \ell_j = \ln \ell(\theta_j)$ where $\theta_j = \mathbf{x}_j \mathbf{b}$, and this provides a shortcut for the calculation of the derivatives. To simplify notation, let us write $f(\theta_j)$ for $\ln \ell(\theta_j)$ and write $f'(\theta_j)$ for its first derivative and $f''(\theta_j)$ for its second. Using the chain rule,

$$\mathbf{g}_j = \frac{\partial f(\theta_j)}{\partial \mathbf{b}} = \frac{\partial f(\theta_j)}{\partial \theta_j} \frac{\partial \theta_j}{\partial \mathbf{b}} = f'(\theta_j)\mathbf{x}_j$$

Thus,

$$\mathbf{g} = \sum_{j=1}^{N} \mathbf{g}_j = \sum_{j=1}^{N} f'(\theta_j)\mathbf{x}_j$$

The k derivatives of the gradient vector can be obtained by calculating just one derivative numerically: $f'(\theta_j)$. Well, in fact we must calculate N numeric derivatives because we need $f'(\theta_j)$ separately evaluated for each observation, but that is not as bad as it sounds. The major cost of evaluating numeric derivatives is the time to evaluate the likelihood function and we can obtain all $f'(\theta_j)$, $j = 1, \ldots, N$, with just two calls to the user-written likelihood-evaluation program, at least if we are willing to use the same h for each observation, and if we assume we already know the optimal h to use.

Similarly, the matrix of second derivatives can be written

$$\mathbf{H} = \sum_{j=1}^{N} f''(\theta_j)\mathbf{x}_j'\mathbf{x}_j$$

and these $k(k+1)/2$ derivatives can similarly be obtained by evaluating one second derivative numerically (requiring two more calls).

This results in a substantial saving of computer time, especially for large values of k.

3.6 The advantages of lf in terms of accuracy

lf is more accurate than d0 and more accurate than d1.

lf is more accurate than d0 for two reasons:

1. lf is more accurate than d0 because the derivatives are obtained by a mix of analytic and numeric derivatives as opposed to method d0's purely numerical calculation. Actually, the way this works out, there is little improvement in the calculation of the gradient vector because the analytic part is linear. The calculation is $\mathbf{g} = \sum f'(\theta_j)\mathbf{x}_j$ with $f'()$ being obtained numerically and the $\mathbf{x}$ being obtained analytically. There is a much greater improvement in the second derivative calculation because nonlinearity is introduced into the function and it is introduced analytically: $\mathbf{H} = \sum f''(\theta_j)\mathbf{x}_j'\mathbf{x}_j$.
2. lf is more accurate than d0 for purely numerical reasons. The calculation is made at the observation level and the result is then summed rather than, as in d0, summing first and then differencing to calculate derivatives.

lf is more accurate than method d1 for a different reason: We wrote better code for lf and invested more in the search for the optimal h.

In addition, reason (1) may still apply to an lf vs. d1 comparison for many problems. In the case of lf, we obtain $\mathbf{H} = \sum f''(\theta_j)\mathbf{x}_j'\mathbf{x}_j$ where $f''()$ is computed numerically but note that $\mathbf{x}_j'\mathbf{x}_j$ is an analytic result. In method d1 we start with an analytic $f'()$ and use that to obtain $f''()$: thus we are obtaining the $\mathbf{x}_j'\mathbf{x}_j$ part numerically and the question becomes whether the numerical calculation of the Hessian from $f()$ in lf introduces more error than the numerical estimation of $\mathbf{x}_j'\mathbf{x}_j$ in the d1 case.

lf is, of course, less accurate than d2, assuming you write the required likelihood, gradient, and Hessian calculation program well.

4 Introduction to methods d0, d1, and d2

Contents

4.1 Overview

Methods d0, d1, and d2 all obtain maximums based on the function, its first derivatives, and its second derivatives. Method d2 requires that you supply analytic calculations for the function, its gradient, and its Hessian. Method d1 requires that you supply analytic calculations for the function and its gradient and substitutes a numerical calculation for the Hessian based on the gradient. Method d0 requires that you supply the function only; it obtains both the gradient and the Hessian numerically.

4.1.1 How methods d0, d1, and d2 differ from method lf

Methods d0, d1, and d2 differ from method lf in what you are passed and what you return:

1. In method lf you are passed the thetas, $\theta_{ij} = \mathbf{x}_{ij}\mathbf{b}_i$, the already evaluated linear combinations.

   Methods d0, d1, and d2 pass you the parameter vectors themselves: $\mathbf{b}_1$, $\mathbf{b}_2$, ... These vectors are combined into a single vector $\mathbf{b} = (\mathbf{b}_1, \mathbf{b}_2, \ldots)$ and it is row vector $\mathbf{b}$ that is passed to your routine. It is your responsibility to obtain the thetas from $\mathbf{b}$. The utility command `mleval`—described below—helps to do that.

2. In method lf you fill in `` `lnf' ``, which is a variable in the dataset, with the observation-by-observation log likelihoods $\ln \ell_j$. Method lf takes responsibility for summing the components to form the overall log likelihood $\ln L$, $\ln L = \ln\ell_1 + \ln\ell_2 + \cdots + \ln\ell_N$.

   Methods d0, d1, and d2 require you to fill in the *scalar* `` `lnf' `` with $\ln L$, the overall log likelihood. The utility command `mlsum`—described below—helps do this.

In addition,

3. You are responsible for working directly with the parameter vector, detecting impossible parameter values (occasions when the likelihood function cannot be evaluated), adjusting calculations for inclusion of weights, and ensuring that calculations are restricted to the estimation subsample.

Method lf quietly handled a lot of problems for you. We provide utility routines `mleval`, `mlsum`, `mlvecsum`, and `mlmatsum` that will continue to handle these problems for you. They are documented in this chapter.

4.1.2 How methods d0, d1, and d2 differ from each other

Methods d0, d1, and d2 differ from each other in that

1. Method d0 requires that you fill in only the overall log likelihood.
2. Method d1 requires that you fill in the overall log likelihood and the overall gradient vector (`mlvecsum` helps do this).
3. Method d2 requires that you fill in the overall log likelihood, the overall gradient vector, and the overall negative Hessian (`mlmatsum` helps do this).

4.1.3 The advantages and disadvantages of methods d0, d1, and d2

There are two reasons why you would use method d0 instead of method lf:

1. Method d0 does not make the assumption that the overall log likelihood $\ln L$ is the sum of the observation-by-observation log likelihoods. Some estimators—such as conditional logistic regression and panel-data models—produce log likelihoods for groups of observations, and so do not lend themselves to method lf implementations.
2. You really want to write a method d2 evaluator and method d0 is a good first step along that path.

You would choose methods d1 and d2 for the same reasons. Method d1 is faster and more accurate than method d0—but it is more work to program—and method d2 is faster and more accurate than that—and yet more work to program.

If your likelihood function can be implemented using method lf, there is no reason to use methods d0 and d1 except as steps along the way to a method d2 evaluator. Method lf is faster and more accurate than either method d0 or d1.

Method d2 is faster and more accurate than method lf, however.

4.1.4 How you specify method d0, d1, or d2

To specify method d0, d1, or d2, you code `d0`, `d1`, or `d2` on the `ml model` line:

```
. ml model d0 myprog ...
```

or

```
. ml model d1 myprog ...
```

or

```
. ml model d2 myprog ...
```

You write *myprog*—your likelihood-function evaluator—according to the d0, d1, or d2 specifications.

4.2 Outline of method d0, d1, and d2 evaluators

Methods d0, d1, and d2 may be used with any likelihood function,

$$
\begin{aligned}
\ln L &= \ln L\big((\theta_{1j}, \theta_{2j}, \ldots, \theta_{Ej}; y_{1j}, y_{2j}, \ldots, y_{Dj}) : j = 1, \ldots, N\big) \\
\theta_{1j} &= \mathbf{x}_{1j}\mathbf{b}_1 \\
\theta_{2j} &= \mathbf{x}_{2j}\mathbf{b}_2 \\
&\vdots \\
\theta_{Ej} &= \mathbf{x}_{Ej}\mathbf{b}_E
\end{aligned}
$$

where j indexes observations.

The outline of the method d0, d1, or d2 evaluator is

```
program define myprog
        version 6
        args todo b lnf g negH
        Form scalar `lnf'=ln L
        if `todo'==0 { exit }              /* method d0 evaluators end here */
        Form row vector `g'=∂ ln L/∂b
        if `todo'==1 { exit }              /* method d1 evaluators end here */
        Form matrix `negH'=−∂² ln L/∂b∂b'
end
```

4.2.1 The todo argument

The first argument a method d0, d1, or d2 evaluator receives is `` `todo' `` and it contains 0, 1, or 2:

0. If `` `todo' `` = 0, the evaluator is to fill in the log likelihood.
1. If `` `todo' `` = 1, the evaluator is to fill in the log likelihood and the gradient vector (stored as a *row* vector).
2. If `` `todo' `` = 2, the evaluator is to fill in the log likelihood, the gradient vector, and the negative of the Hessian.

Method d0 evaluators will receive `` `todo' `` = 0 only.

Method d1 evaluators will receive `` `todo' `` = 0 or `` `todo' `` = 1. When called with `` `todo' `` = 0, the evaluator may calculate **g**, but it need not. If calculated, it is ignored.

Method d2 evaluators will receive `` `todo' `` = 0, `` `todo' `` = 1, or `` `todo' `` = 2. Regardless of the value received, the evaluator may calculate more than is requested and the extra is ignored.

4.2.2 The parameter vector argument

The second argument a method d0, d1, or d2 evaluator receives is `` `b' `` and it contains the entire parameter vector as a row vector. For instance, were you to type

```
. ml model d0 myprog (foreign=mpg weight)
```

when you define your estimation problem, then the parameter vector `b´ that your program would later receive would contain three elements: the coefficient on mpg, the coefficient on weight, and the intercept. The interpretation would be

$$\theta_{1j} = b_1\texttt{mpg}_j + b_2\texttt{weight}_j + b_3$$

Were you to type

```
. ml model d0 myprog (foreign=mpg weight) (displ=weight)
```

then vector `b´ would contain five elements representing two equations:

$$\theta_{1j} = b_1\texttt{mpg}_j + b_2\texttt{weight}_j + b_3$$
$$\theta_{2j} = b_4\texttt{weight}_j + b_5$$

Were you to type

```
. ml model d0 myprog (foreign=mpg weight) (displ=weight) /sigma
```

then vector `b´ would contain six elements representing three equations:

$$\theta_{1j} = b_1\texttt{mpg}_j + b_2\texttt{weight}_j + b_3$$
$$\theta_{2j} = b_4\texttt{weight}_j + b_5$$
$$\theta_{3j} = b_6$$

The equation order, and the order of coefficients within equation, is exactly as you specify them on the ml model statement. In addition, ml sets the column names of the parameter vector you are passed with the equation and variable names.

When you typed

```
. ml model d0 myprog (foreign=mpg weight)
```

were your program called with `b´ $= (1, 2, 3)$, the parameter vector your program would receive would be

```
          eq1:    eq1:    eq1:
          mpg  weight   _cons
r1          1       2       3
```

In the final example, where you typed

```
. ml model d0 myprog (foreign=mpg weight) (displ=weight) /sigma
```

were the parameter vector `b´ $= (1, 2, 3, 4, 5, 6)$, your program would receive

```
          eq1:    eq1:    eq1:    eq2:    eq2:  sigma:
          mpg  weight   _cons  weight   _cons   _cons
r1          1       2       3       4       5       6
```

4.2.3 Using mleval to obtain thetas from the parameter vector

The utility `mleval` calculates thetas from the parameter vector. The syntax of `mleval` is

`mleval` *newvarname* = *vecname* [, `eq(`*#*`)`]

vecname refers to the parameter vector passed by `ml` (i.e., `` `b' ``). *newvarname*, stored as a `double`, is created containing θ_{ij} corresponding to the ith equation. Option `eq()` specifies i, the equation to be evaluated; it defaults to `eq(1)`.

`mleval` is typically used as follows,

```
program define progname
        version 6
        args todo b lnf g negH
        tempvar theta1 theta2 ...
        mleval `theta1' = `b', eq(1)
        mleval `theta2' = `b', eq(2)
        ...
end
```

The remainder of the program can then be written in terms of `` `theta1' ``, `` `theta2' ``, ...

For instance, pretend the `ml model` statement you previously typed is

```
. ml model d0 probit (foreign=mpg weight)
```

then your program might read

```
program define probit
        version 6
        args todo b lnf
        tempvar theta
        mleval `theta' = `b'
        ...
end
```

Now let's consider another model. You type

```
. ml model d2 xmpl2 (foreign=mpg weight) (displ=weight) /sigma
```

and in that case you might code

```
program define xmpl2
        version 6
        args todo b lnf g negH
        tempvar theta1 theta2 theta3
        mleval `theta1' = `b', eq(1)
        mleval `theta2' = `b', eq(2)
        mleval `theta3' = `b', eq(3)
        ...
end
```

In this example, note that θ_3 is a constant. Thus, there would be an efficiency gain to making `` `theta3' `` a scalar rather than a variable in the dataset. `mleval` has a second syntax for such instances:

`mleval` *scalarname* = *vecname*, `scalar` [`eq(`*#*`)`]

When you specify the `scalar` option, `mleval` creates a scalar:

```
program define xmpl3
        version 6
        args todo b lnf g negH
        tempvar theta1 theta2
        tempname theta3
        mleval `theta1' = `b', eq(1)
        mleval `theta2' = `b', eq(2)
        mleval `theta3' = `b', eq(3) scalar
        ...
end
```

If you were to attempt to use the `scalar` option on an inappropriate equation (such as `eq(1)` in the example above), `mleval` will issue the error "mleval, scalar: eq(1) not constant", r(198).

When evaluating equations, use `mleval` without the `scalar` option for equations that produce values that vary across observations or that might vary across observations. Optionally specify the `scalar` option for equations that produce values that are constant. Specifying `scalar` will produce slightly more efficient code.

4.2.4 The lnf argument

The third argument a method d0, d1, or d2 evaluator receives is `` `lnf' ``, the name of a scalar you are to create containing the log-likelihood value evaluated at `` `b' ``. If the log likelihood cannot be calculated, `` `lnf' `` is to contain missing.

To aid understanding, here is how you might code the probit estimator:

```
program define myprobit
        version 6
        args todo b lnf
        tempvar theta lnfj
        mleval `theta' = `b'
        quietly gen double `lnfj' = ln(normprob(`theta')) if $ML_y1==1
        quietly replace `lnfj' = ln(1-normprob(`theta')) if $ML_y1==0
        quietly replace `lnfj' = sum(`lnfj')
        scalar `lnf' = `lnfj'[_N]
end
```

Do not take this example too seriously; it has a lot of problems:

1. When you use methods d0, d1, or d2, it is your responsibility to restrict your calculations to the estimation subsample. Note how we sum over all the observations, including, possibly, observations that should be excluded. Perhaps the user typed

   ```
   . ml model d0 myprobit (for=mpg weight) if mpg>20
   ```

   Do not make too much of this problem because it is your responsibility to restrict calculations to the estimation subsample only when you specify `ml model`'s `nopreserve` option. Otherwise, `ml` automatically saves your data and then drops all the irrelevant observations, so you can be sloppy. (`ml` does not do this with method lf because method lf sums the observations and it does so carefully.) Nevertheless, preserving and restoring the data takes time and you might be tempted to specify `ml model`'s `nopreserve` option. With the `nopreserve` option, the program above would then produce incorrect results.

2. We did nothing to account for weights, should they be specified. That will not be a problem as long as no weights are specified on the `ml model` statement, but a better draft of this program would replace

   ```
           quietly replace `lnfj' = sum(`lnfj')
   ```

with

```
quietly replace `lnfj' = sum($ML_w*`lnfj')
```

Using methods d0, d1, or d2, weights are your responsibility.

3. We merely assumed that the log likelihood can be calculated for every observation. Let's pretend, however, that the values we received for `` `b' `` are odd and result in the calculation of the log likelihood being missing in the third observation. The `sum()` function used at the last step will treat the missing value as contributing 0 and so calculate an overall log-likelihood value that will be incorrect. Worse, treating $\ln \ell_3 = 0$ is tantamount to $\ell_3 = 1$, and in this case, because likelihoods correspond to probabilities, the impossible value is treated as producing a probability 1 result.

   You may think this unlikely, but it is not. We direct your attention to line 6 of our program:

   ```
   quietly replace `lnfj' = ln(1-normprob(`theta')) if $ML_y1==0
   ```

   `` ln(1-normprob(`theta')) `` turns out to be a very poor way of calculating $\ln \ell$ when the observed outcome is 0. `ln(1-normprob(9))`, for instance, evaluates to missing! A better way is `` ln(normprob(-`theta')) `` and `ln(normprob(-9))` is −43.63. Now we should not use a poor numerical calculation formula, but the fact that `` ln(1-normprob(`theta')) `` is poor may not have been obvious to you. Method lf protected you by watching for when things went badly; methods d0, d1, and d2 do not because they cannot—they never see the observation-by-observations results and indeed, for some method d0, d1, and d2 evaluators, there may be no such thing.

This third problem is potentially serious and therefore we strongly urge you to avoid Stata's `sum()` function and use the `mlsum` command—described below—to fill in the log likelihood. Using `mlsum`, an adequate version of our probit subroutine would read

```
program define myprobit
        version 6
        args todo b lnf
        tempvar theta lnfj
        mleval `theta' = `b'
        quietly gen double `lnfj' = ln(normprob(`theta')) if $ML_y1==1
        quietly replace `lnfj' = ln(1-normprob(`theta')) if $ML_y1==0
        mlsum `lnf' = `lnfj'
end
```

`mlsum` addresses all three of the shortcomings we have mentioned.

4.2.5 Using lnf to indicate that the likelihood cannot be calculated

Your evaluator might be called with values of `` `b' `` for which the log likelihood cannot be evaluated for substantive as well as numerical reasons; for instance, perhaps one of the parameters to be estimated is a variance and the maximizer attempts to evaluate the function with this parameter set to a negative value.

If the likelihood function cannot be evaluated at the particular parameter values `` `b' ``, your subroutine is to return missing in `` `lnf' ``. It need not fill in `` `g' `` and `` `negH' `` even if `` `todo' `` would ordinarily call for such a calculation, although it may fill them in even so. While trash values of `` `g' `` and `` `negH' `` will not bother `ml` in such instances, if the gradient or negative Hessian would themselves contain missing values, that will bother Stata. Stata does not allow vectors and matrices to contain missing values and will issue the error message "matrix has missing values"; r(109). Thus, after setting `` `lnf' `` to missing, it is best to exit and thus skip the remaining calculations, if any.

The issue of not being able to calculate the log likelihood never arose with method lf because lf could detect the problem by examining what your subroutine returned. With method lf, you calculated observation-by-observation log-likelihood values. If some of the returned values turned out to be missing—whether for numerical or substantive reasons—lf could spot the missing values and take the appropriate action.

Methods d0, d1, and d2 cannot do this because they never see the observation-by-observation values. The evaluator you write returns the overall log likelihood. It is therefore your responsibility to watch for impossible parameter values and let the maximizer know. If you encounter impossible values, set `` `lnf' `` to contain missing and exit.

4.2.6 Using mlsum to define lnf

Most likelihood-evaluation programs generate a log likelihood for each observation or group of observations and then sum the contributions to obtain the overall log-likelihood value. `mlsum` will perform the summation. The syntax of `mlsum` is

`mlsum` *scalarname*$_{\rm lnf}$ = *exp* [`if` *exp*]

There are other ways you could make this sum in Stata, but there are three reasons why you should use `mlsum`:

1. `mlsum` automatically restricts the summation to the estimation subsample (observations for which `$ML_samp==1`) and thus makes it more likely that results are correct even if you specify `ml model`'s `nopreserve` option.
2. `mlsum` automatically applies weights, if they are specified, in forming the sum.
3. `mlsum` verifies that what is being summed contains no missing values or it sets *scalarname*$_{\rm lnf}$ to contain missing if it does.

`mlsum` is typically used as follows:

```
program define myll
        version 6
        args todo b lnf g negH
        ...
        mlsum `lnf' = ...
        if `lnf'==. { exit }
        ...
end
```

The right-hand side of `mlsum` is filled in with something that, when summed across the observations, yields the overall log-likelihood value. The above outline is relevant when log-likelihood values are calculated for every observation in the estimation subsample. `mlsum` itself will handle restricting the summation to the relevant observations (the observations for which `$ML_samp==1`).

When log-likelihood values exist for only groups of observations, the outline is

```
program define myll
        version 6
        args todo b lnf g negH
        ...
        mlsum `lnf' = ... if ...
        if `lnf'==. { exit }
        ...
end
```

The `if` *exp* must be filled in to be true for observations where you expect log-likelihood values to exist. `mlsum` does *not* skip over missing values in the estimation subsample. It takes missing as an indication that the log-likelihood value could not be calculated (and so sets the `` `lnf' `` also to contain missing).

For instance, the log likelihood for probit is

$$\ln L = \sum_{j=1}^{N} \ln \ell_j$$

$$\ln \ell_j = \begin{cases} \ln \Phi(\theta_j) & \text{if } y_j = 1 \\ \ln \Phi(-\theta_j) & \text{if } y_j = 0 \end{cases}$$

$$\theta_j = \mathbf{x}_j \mathbf{b}$$

The dots in `mlsum` need to be filled in with the formula for $\ln \ell_j$:

```
tempvar lnfj
gen double `lnfj' = ln(normprob(`theta')) if $ML_y1==1
replace `lnfj' = ln(normprob(-`theta')) if $ML_y1==0
mlsum `lnf' = `lnfj'
```

Note that we can specify an expression following `mlsum`, so the above four lines of code could be more compactly coded

```
mlsum `lnf' = ln(normprob(cond($ML_y1,`theta',-`theta')))
```

4.2.7 The g argument

The fourth argument a method d0, d1, or d2 evaluator receives is `` `g' ``, the name of a matrix (vector) you are to create containing the gradient. You need only create `` `g' `` if `` `todo' `` is 1 or 2 (which would happen, of course, only if you are using methods d1 or d2).

Method d0 evaluators never need to fill in `` `g' `` and, in fact, most people would code a method d0 evaluator as

```
program define myprog
        version 6
        args todo b lnf
        Form scalar `lnf'=ln L
end
```

In fact, however, even method d0 evaluators receive third argument `` `g' `` (and fourth argument `` `negH' ``). It is just that they are called only with `` `todo' `` = 0, and so they can ignore the third and fourth arguments. This feature is useful because it allows a method d1 or d2 evaluator to be used with method d0. This is useful in debugging and checking results. Say you have a method d2 evaluator you wrote some time ago and you obtain an odd result and so wonder if it is really calculating the first and second derivatives correctly in this case. If you typed

```
. ml model d2 xmpl2 (foreign=mpg weight) (displ=weight) /sigma
. ml maximize
```

and obtained odd results, you can type

```
. ml model d0 xmpl2 (foreign=mpg weight) (displ=weight) /sigma
. ml maximize
```

to obtain method d0 results, substituting numerical calculations for the suspect first and second derivatives. Note that you make no changes whatsoever to your program.

In any case, a method d1 or method d2 evaluator will be called upon to fill in `` `g' ``. The first derivatives (gradient) are to be stored as a row vector that is conformable with the `` `b' `` vector you were passed. This means that if `` `b' `` is $1 \times k$, `` `g' `` is to be $1 \times k$ and it is to have the derivatives in the same order as the parameters in `` `b' ``.

`` `g' `` is to contain the partial derivatives of $\ln L$ with respect to each of the parameters:

$$\mathbf{g} = \left(\frac{\partial \ln L}{\partial b_1}, \frac{\partial \ln L}{\partial b_2}, \ldots, \frac{\partial \ln L}{\partial b_k} \right)$$

The row and column names placed on `` `g' `` do not matter because they are ignored by `ml`.

If `` `g' `` cannot be calculated at the current values of the parameter vector `` `b' ``, `` `lnf' `` is to be set to contain missing and the contents of `` `g' ``, or even whether `` `g' `` is defined, do not matter.

4.2.8 Using mlvecsum to define g

`mlvecsum` assists in calculating the gradient vector when the linear-form restrictions are met. In that case,

$$\begin{aligned}
\ln L &= \sum_{j=1}^{N} \ln \ell(\theta_{1j}, \theta_{2j}, \ldots, \theta_{Ej}) \\
\theta_{1j} &= \mathbf{x}_{1j}\mathbf{b}_1 \\
\theta_{2j} &= \mathbf{x}_{2j}\mathbf{b}_2 \\
&\ldots \\
\theta_{Ej} &= \mathbf{x}_{Ej}\mathbf{b}_E
\end{aligned}$$

and $\mathbf{b} = (\mathbf{b}_1, \mathbf{b}_2, \ldots, \mathbf{b}_E)$. Thus, the gradient vector can be written

$$\begin{aligned}
\frac{\partial \ln L}{\partial \mathbf{b}} &= \left(\sum_{j=1}^{N} \frac{\partial \ln \ell_j}{\partial \theta_{1j}} \frac{\partial \theta_{1j}}{\partial \mathbf{b}_1}, \sum_{j=1}^{N} \frac{\partial \ln \ell_j}{\partial \theta_{2j}} \frac{\partial \theta_{2j}}{\partial \mathbf{b}_2}, \ldots, \sum_{j=1}^{N} \frac{\partial \ln \ell_j}{\partial \theta_{Ej}} \frac{\partial \theta_{Ej}}{\partial \mathbf{b}_E} \right) \\
&= \left(\sum_{j=1}^{N} \frac{\partial \ln \ell_j}{\partial \theta_{1j}} \mathbf{x}_{1j}, \sum_{j=1}^{N} \frac{\partial \ln \ell_j}{\partial \theta_{2j}} \mathbf{x}_{2j}, \ldots, \sum_{j=1}^{N} \frac{\partial \ln \ell_j}{\partial \theta_{Ej}} \mathbf{x}_{Ej} \right)
\end{aligned}$$

$\partial \ln \ell_j / \partial \theta_{ij}$ is simply one value per observation. You supply $\partial \ln \ell_j / \partial \theta_{ij}$ and `mlvecsum` returns $\sum_j (\partial \ln \ell_j / \partial \theta_{ij}) \mathbf{x}_{ij} = \partial \ln L / \partial \mathbf{b}_i$, which is one component of the vector above. `mlvecsum`'s syntax is

`mlvecsum` *scalarname*$_{\text{lnf}}$ *rowvecname* = *exp* [`if` *exp*] [, `eq(`*#*`)`]

where *exp* evaluates to $\partial \ln \ell_j / \partial \theta_{ij}$ for equation $i = \#$. Thus, obtaining $\partial \ln L / \partial \mathbf{b}$ is simply a matter of issuing one `mlvecsum` per equation and then joining the results into a single row vector.

In a single-equation system:

```
mlvecsum `lnf' `g' = formula for ∂ln ℓj/∂θ1j
```

In a two-equation system:

```
tempname dt1 dt2
mlvecsum `lnf' `dt1' = formula for ∂ln ℓj/∂θ1j, eq(1)
mlvecsum `lnf' `dt2' = formula for ∂ln ℓj/∂θ2j, eq(2)
matrix `g' = (`dt1',`dt2')
```

In a three-equation system:

```
tempname dt1 dt2 dt3
mlvecsum `lnf' `dt1' = formula for ∂ln ℓj/∂θ1j, eq(1)
mlvecsum `lnf' `dt2' = formula for ∂ln ℓj/∂θ2j, eq(2)
mlvecsum `lnf' `dt3' = formula for ∂ln ℓj/∂θ3j, eq(3)
matrix `g' = (`dt1',`dt2',`dt3')
```

and so on.

Note that `mlvecsum` has two left-hand-side arguments:

```
. mlvecsum `lnf' `whatever' = exp
```

That is so `mlvecsum` can reset `` `lnf' `` to contain missing if *exp* cannot be calculated for some relevant observation. Method d0, d1, and d2 evaluators are to set `` `lnf' `` to missing when called with impossible values of the parameter vector. It is possible that `` `lnf' `` can be calculated but the gradient cannot. `mlvecsum` appropriately handles such cases.

As an example, the log likelihood for probit is

$$\ln L = \sum_{j=1}^{N} \ln \ell_j$$

$$\ln \ell_j = \begin{cases} \ln \Phi(\theta_j) & \text{if } y_j = 1 \\ \ln \Phi(-\theta_j) & \text{if } y_j = 0 \end{cases}$$

$$\theta_j = \mathbf{x}_j \mathbf{b}$$

Probit has only one equation (one theta). The observation-by-observation derivative is

$$\frac{d \ln \ell_j}{d\theta_j} = \begin{cases} \phi(\theta_j)/\Phi(\theta_j) & \text{if } y_j = 1 \\ -\phi(\theta_j)/\Phi(-\theta_j) & \text{if } y_j = 0 \end{cases}$$

Thus, `mlvecsum` could be used to obtain the entire gradient vector by coding

```
tempvar gj
gen double `gj' = normd(`theta')/normprob(`theta') if $ML_y1==1
replace `gj' = -normd(`theta')/(normprob(-`theta')) if $ML_y1==0
mlvecsum `lnf' `g' = `gj'
```

Note that because the right-hand side of `mlvecsum` can be an expression, the above could be reduced to one line using the `cond()` function if we preferred.

4.2.9 The negH argument

The fifth argument a method d0, d1, or d2 evaluator receives is `` `negH' ``, the name of a matrix you are to create containing the negative of the Hessian (negative of the second derivatives). Only d2 evaluators need to create `` `negH' `` and then only when `` `todo' `` is 2, although they may create it at other times, too. (Method d0 and d1 evaluators may calculate `` `negH' `` if they wish, but it will be ignored, and all evaluators may fill in `` `negH' `` even when not called upon to do so.)

The negative second derivatives are to be stored as a matrix that is conformable with the `` `b' `` vector you were passed. This means that if `` `b' `` is $1 \times k$, `` `negH' `` is to be $k \times k$, and it is to have the derivatives in the same order as the parameters in `` `b' ``.

`` `negH' `` is to contain the negative second partial derivatives of $\ln L$ with respect to each of the parameters:

$$
-\mathbf{H} = \begin{pmatrix}
-\frac{\partial \ln L}{\partial b_1^2} & -\frac{\partial \ln L}{\partial b_1 \partial b_2} & \cdots & -\frac{\partial \ln L}{\partial b_1 \partial b_k} \\
-\frac{\partial \ln L}{\partial b_2 \partial b_1} & -\frac{\partial \ln L}{\partial b_2^2} & \cdots & -\frac{\partial \ln L}{\partial b_2 \partial b_k} \\
\vdots & \vdots & \ddots & \vdots \\
-\frac{\partial \ln L}{\partial b_k \partial b_1} & -\frac{\partial \ln L}{\partial b_k \partial b_2} & \cdots & -\frac{\partial \ln L}{\partial b_k^2}
\end{pmatrix}
$$

The row and column names placed on `` `negH' `` do not matter because they are ignored by `ml`.

If `` `negH' `` cannot be calculated at the current values of the parameter vector `` `b' ``, `` `lnf' `` is to be set to contain missing and the contents of `` `negH' ``, or even whether it is defined, do not matter.

4.2.10 Using mlmatsum to define negH

`mlmatsum` works like `mlvecsum`—it assists in calculating the negative Hessian matrix when the linear-form restrictions are met. In that case the negative Hessian can be written

$$
-\frac{\partial^2 \ln \ell_j}{\partial \mathbf{b} \partial \mathbf{b}'} = \begin{pmatrix}
\sum_j -\frac{\partial^2 \ln \ell_j}{\partial \theta_{1j}^2} \mathbf{x}_{1j}' \mathbf{x}_{1j} & \sum_j -\frac{\partial^2 \ln \ell_j}{\partial \theta_{1j} \partial \theta_{2j}} \mathbf{x}_{1j}' \mathbf{x}_{2j} & \cdots & \sum_j -\frac{\partial^2 \ln \ell_j}{\partial \theta_{1j} \partial \theta_{kj}} \mathbf{x}_{1j}' \mathbf{x}_{kj} \\
\sum_j -\frac{\partial^2 \ln \ell_j}{\partial \theta_{2j} \partial \theta_{1j}} \mathbf{x}_{2j}' \mathbf{x}_{1j} & \sum_j -\frac{\partial^2 \ln \ell_j}{\partial \theta_{2j}^2} \mathbf{x}_{2j}' \mathbf{x}_{2j} & \cdots & \sum_j -\frac{\partial^2 \ln \ell_j}{\partial \theta_{2j} \partial \theta_{kj}} \mathbf{x}_{2j}' \mathbf{x}_{kj} \\
\vdots & \vdots & \ddots & \vdots \\
\sum_j -\frac{\partial^2 \ln \ell_j}{\partial \theta_{kj} \partial \theta_{1j}} \mathbf{x}_{kj}' \mathbf{x}_{1j} & \sum_j -\frac{\partial^2 \ln \ell_j}{\partial \theta_{kj} \partial \theta_{2j}} \mathbf{x}_{kj}' \mathbf{x}_{2j} & \cdots & \sum_j -\frac{\partial^2 \ln \ell_j}{\partial \theta_{kj}^2} \mathbf{x}_{kj}' \mathbf{x}_{kj}
\end{pmatrix}
$$

As with `mlvecsum`, $\partial^2 \ln \ell_j / \partial \theta_{ij} \partial \theta_{kj}$ is just a value per observation; you supply $\partial^2 \ln \ell_j / \partial \theta_{ij} \partial \theta_{kj}$ and `mlmatsum` returns $\sum_j (\partial^2 \ln \ell_j / \partial \theta_{ij} \partial \theta_{kj}) \mathbf{x}_{ij}' \mathbf{x}_{kj}$, which is one element (itself a matrix) of the above matrix. Its syntax is

`mlmatsum` *scalarname*$_{\text{lnf}}$ *matrixname* = *exp* [`if` *exp*] [, `eq(`#[,#]`)`]

Thus, obtaining $\partial^2 \ln L / \partial \mathbf{b} \partial \mathbf{b}'$ is simply a matter of issuing one `mlmatsum` per equation pair and joining the results into a single matrix. For instance,

In a single-equation system:

mlmatsum `` `lnf' `negH' `` = formula for $-\partial^2 \ln \ell_j / \partial \theta_{1j}^2$

In a two-equation system:

```
tempname d11 d12 d22
mlmatsum `lnf' `d11' = formula for −∂²ln ℓj/∂θ²1j, eq(1,1)
mlmatsum `lnf' `d12' = formula for −∂²ln ℓj/∂θ1j∂θ2j, eq(1,2)
mlmatsum `lnf' `d22' = formula for −∂²ln ℓj/∂θ²2j, eq(2,2)
matrix `negH' = (`d11',`d12' \ `d12'',`d22')
```

and so on.

Note that `mlmatsum` has two left-hand-side arguments:

```
. mlmatsum `lnf' `whatever' = exp
```

That is so `mlmatsum` can reset `` `lnf' `` to contain missing if *exp* cannot be calculated for some relevant observation. Method d0, d1, and d2 evaluators are to set `` `lnf' `` to missing when called with impossible values of the parameter vector. It is possible that `` `lnf' `` can be calculated but the negative Hessian cannot. `mlmatsum` appropriately handles such cases.

For example, the log likelihood for probit is

$$\ln L = \sum_{j=1}^{N} \ln \ell_j$$

$$\ln \ell_j = \begin{cases} \ln \Phi(\theta_j) & \text{if } y_j = 1 \\ \ln \Phi(-\theta_j) & \text{if } y_j = 0 \end{cases}$$

$$\theta_j = \mathbf{x}_j \mathbf{b}$$

and, as we have noted previously, probit has one equation and therefore one theta, and the observation-by-observation first derivative is

$$\frac{d \ln \ell_j}{d\theta_j} = \begin{cases} \phi(\theta_j)/\Phi(\theta_j) & \text{if } y_j = 1 \\ -\phi(\theta_j)/\Phi(-\theta_j) & \text{if } y_j = 0 \end{cases}$$

The observation-by-observation negative second derivative is

$$-\frac{d^2 \ln \ell_j}{d\theta_j^2} = \begin{cases} R(\theta_j)\big(R(\theta_j) + \theta_j\big) & \text{if } y_j = 1 \\ S(\theta_j)\big(S(\theta_j) - \theta_j\big) & \text{if } y_j = 0 \end{cases}$$

$$R(\theta_j) = \phi(\theta_j)/\Phi(\theta_j)$$

$$S(\theta_j) = \phi(\theta_j)/\Phi(-\theta_j)$$

Thus, `mlmatsum` could be used to obtain the entire negative Hessian by coding

```
tempvar R S h
gen double `R' = normd(`theta')/normprob(`theta')
gen double `S' = normd(`theta')/normprob(-`theta')
gen double `h' = `R'*(`R'+`theta') if $ML_y1==1
replace    `h' = `S'*(`S'-`theta') if $ML_y1==0
mlmatsum `lnf' `negH' = `h'
```

Note that because the right-hand side of `mlmatsum` can be an expression, the above could be simplified to fewer lines using the `cond()` function if we preferred.

4.3 pweights and robust estimates of variance

`ml` can produce robust estimates of variance *when the linear-form restrictions are met*; that is,

$$\ln L = \ln \ell_1 + \ln \ell_2 + \cdots + \ln \ell_N$$

which is to say, when the observations are independent, or independent at least as far as the likelihood calculation is concerned. There must be a log-likelihood value for each observation in the estimation subsample and those contributions must sum to the overall log likelihood.

In order to produce robust variance estimates, `ml` needs to know the observation-by-observation scores, $\partial \ell_j / \partial \theta_{ij}$, for each of the equations, $i = 1, 2, \ldots, E$. These scores are exactly what are used by `mlvecsum` to produce the gradient vector:

`mlvecsum` *scalarname*$_{\mathrm{lnf}}$ *rowvecname* = *exp* [`if` *exp*] [, `eq(`*#*`)`]

where *exp* evaluates to $\partial \ln \ell_j / \partial \theta_{ij}$. In the case of probit, for instance, these scores are

$$\frac{d\ln \ell_j}{d\theta_j} = \begin{cases} \phi(\theta_j)/\Phi(\theta_j) & \text{if } y_j = 1 \\ -\phi(\theta_j)/\Phi(-\theta_j) & \text{if } y_j = 0 \end{cases}$$

and, in **4.2.8 Using mlvecsum**, we offered the following for calculating the gradient vector:

```
tempvar gj
gen double `gj´ = normd(`theta´)/normprob(`theta´) if $ML_y1==1
replace `gj´ = -normd(`theta´)/(normprob(-`theta´)) if $ML_y1==0
mlvecsum `lnf´ `g´ = `gj´
```

The temporary variable `` `gj´ `` in the above is exactly the score that `ml` needs to calculate the robust estimate of variance.

We have said that `ml` passes d0, d1, and d2 evaluators five arguments:

1. `` `todo´ ``, equal to 0, 1, or 2.
2. `` `b´ ``, a coefficient vector.
3. `` `lnf´ ``, a scalar to receive $\ln L$.
4. `` `g´ ``, a vector to receive $\partial \ln L / \partial \mathbf{b}$.
5. `` `negH´ ``, a matrix to receive $-\partial^2 \ln L / \partial \mathbf{b} \partial \mathbf{b}'$.

In fact, `ml` passes $5 + E$ arguments, the remaining E being

6. `` `g1´ ``, a variable to receive $\partial \ln \ell / \partial \theta_1$.
7. `` `g2´ ``, a variable to receive $\partial \ln \ell / \partial \theta_2$, if there is a second equation.
8. ...

The variables `` `g1´ ``, `` `g2´ ``, ..., exist (they are `doubles`) and are filled in with missing values. You can leave them that way. For some problems `` `g1´ ``, `` `g2´ ``, ..., make no sense because there is no concept of an observation-by-observation value of the score—the likelihood function does not meet the linear-form restrictions.

When the likelihood function does meet the linear-form restrictions, however, and when you want to use `ml model`'s `robust` or `cluster()` options, or you want to specify `pweights`, you need to fill in these variables.

Filling them in is easy: you have already calculated the quantities needed because **mlvecsum** also needed them. It is simply a matter of storing the values in the right place. In the case of probit, for instance, rather than coding

```
tempvar gj
gen double `gj´ = normd(`theta´)/normprob(`theta´) if $ML_y1==1
replace `gj´ = -normd(`theta´)/(normprob(-`theta´)) if $ML_y1==0
mlvecsum `lnf´ `g´ = `gj´
```

better would be to code

```
replace `g1´ = normd(`theta´)/normprob(`theta´) if $ML_y1==1
replace `g1´ = -normd(`theta´)/(normprob(-`theta´)) if $ML_y1==0
mlvecsum `lnf´ `g´ = `g1´
```

and then to change the **args** statement at the top of the program from

```
args todo b lnf g negH
```

to

```
args todo b lnf g negH g1
```

5 Method d0

Contents

5.1 Basic syntax

Method d0 is for use (1) when the likelihood function violates the linear-form restrictions or (2) as a first step in implementing a method d1 or d2 evaluator. Method d0 may be used with any likelihood function,

$$
\begin{aligned}
\ln L &= \ln L\big((\theta_{1j}, \theta_{2j}, \ldots, \theta_{Ej}; y_{1j}, y_{2j}, \ldots, y_{Dj}) : j = 1, \ldots, N\big) \\
\theta_{1j} &= \mathbf{x}_{1j}\mathbf{b}_1 \\
\theta_{2j} &= \mathbf{x}_{2j}\mathbf{b}_2 \\
&\vdots \\
\theta_{Ej} &= \mathbf{x}_{Ej}\mathbf{b}_E
\end{aligned}
$$

where j indexes observations.

The method d0 evaluator is to calculate $\ln L$, the overall log likelihood. The outline for the program is

```
program define myprog
        version 6
        args todo b lnf                         /* `todo' is subsequently ignored */
        tempvar theta1 theta2 ...
        mleval `theta1' = `b'
        mleval `theta2' = `b', eq(2) /* if there is a θ2 */
        ...
        /* if you need to create any intermediate results: */
        tempvar tmp1 tmp2 ...
        gen double `tmp1' = ...
        ...
        mlsum `lnf' = ...
end
```

1. You are passed the parameter vector. Use `mleval` to obtain the thetas from it; see **4.2.3 Using mleval**.
2. Access the dependent variables (if any) by typing `$ML_y1`, `$ML_y2`, ... Use `$ML_y1`, `$ML_y2`, ..., just as you would any existing variable in the data.

3. Final results are to be saved in scalar `` `lnf' ``. Use `mlsum` to produce it; see **4.2.6 Using mlsum**.
4. It is of great importance that, if you create any temporary variables for intermediate results, you create them as `doubles`: Type "`gen double` *name* `=` ..."; do not omit the word `double`.
5. The estimation subsample is `$ML_samp==1`. You may safely ignore this if you do not specify `ml model`'s `nopreserve` option. When your program is called, only relevant data will be in memory. If you do specify `nopreserve`, understand that `mleval` and `mlsum` automatically restrict themselves to the estimation subsample; it is not necessary to code `if $ML_samp==1` on these commands. You need to restrict other calculation commands to the `$ML_samp==1` subsample.
6. The weights are stored in variable `$ML_w`, which contains 1 in every observation if no weights are specified. If you use `mlsum` to produce the log likelihood, you may ignore this because `mlsum` handles that itself.
7. While method d0 evaluators can be used with `fweights`, `aweights`, and `iweights`, `pweights` may not be specified. Use method lf or method d1 if this is important.
8. You may not specify `ml model`'s `robust` or `cluster()` options with method d0 evaluators. Use method lf or method d1 if this is important.

You specify that you are using method d0 on the `ml model` statement,

```
. ml model d0 myprog ...
```

which you issue after you have defined the likelihood-evaluation program.

5.1.1 Using method d0

Using method d0 is a matter of writing down the likelihood function and coding it. The coding, however, is more work than in the case of method lf because you must evaluate the coefficient vector to obtain the thetas and you must sum the likelihood to produce the overall log-likelihood value.

Method d0, d1, and d2 evaluators all receive the same five arguments:

1st argument:	0, 1, or 2. 0 means calculate $\ln L$ value only. 1 means calculate $\ln L$ value and gradient vector. 2 means calculate $\ln L$ value, gradient vector, and negative Hessian matrix.
2nd argument:	name of coefficient vector.
3rd argument:	name of scalar to be filled in with $\ln L$ value.
4th argument:	name of vector to be filled in with gradient.
5th argument:	name of matrix to be filled in with negative Hessian.

Method d0 routines will be called with 1st argument `0` only, so method d0 evaluators ignore `` `todo' `` and generally ignore that they receive arguments 4 and 5. That is, formally, a method d0 evaluator should open

```
program define myprog
        version 6
        args todo b lnf g negH
        ...
end
```

but it is perfectly acceptable to code

```
program define myprog
        version 6
        args todo b lnf
        ...
end
```

Since the likelihood function is

$$\begin{aligned}\ln L &= \ln L\big((\theta_{1j}, \theta_{2j}, \ldots, \theta_{Ej}; y_{1j}, y_{2j}, \ldots, y_{Dj}) : j = 1, \ldots, N\big)\\ \theta_{1j} &= \mathbf{x}_{1j}\mathbf{b}_1\\ \theta_{2j} &= \mathbf{x}_{2j}\mathbf{b}_2\\ &\vdots\\ \theta_{Ej} &= \mathbf{x}_{Ej}\mathbf{b}_E\end{aligned}$$

the first step is to obtain the theta values from `` `b' ``,

```
program define myprog
        version 6
        args todo b lnf
        tempvar theta1 theta2 ...
        mleval `theta1' = `b'
        mleval `theta2' = `b', eq(2) /* if there is a θ2 */
        ...
end
```

and the final step is to calculate the overall log-likelihood value and store it in the scalar `` `lnf' ``. Most likelihood functions produce a value for each observation or clump of observations. If the function produces a value for each observation, and the function is simple, then the entire code can read

```
program define myprog
        version 6
        args todo b lnf
        tempvar theta1 theta2 ...
        mleval `theta1' = `b'
        mleval `theta2' = `b', eq(2) /* if there is a θ2 */
        ...
        mlsum `lnf' = ...
end
```

where the dots following **mlsum** are filled in with an expression producing the log-likelihood value for an observation, which will be in terms of **`theta1'**, **`theta2'**, ..., **$ML_y1**, **$ML_y2**, ...

In more complicated cases, you need to generate temporary variables containing partial calculations before invoking **mlsum** to sum the contributions:

```
program define myprog
        version 6
        args todo b lnf
        tempvar theta1 theta2 ...
        mleval `theta1' = `b'
        mleval `theta2' = `b', eq(2) /* if there is a θ2 */
        ...
        tempvar tmp1 tmp2
        gen double `tmp1' = ...
        gen double `tmp2' = ...
        mlsum `lnf' = expression in terms of `tmp1' and `tmp2'
end
```

5.1.2 Example 1: Probit

Many likelihood functions meet the linear-form requirements and they are particularly easy to code. You calculate the observation-by-observation likelihood values ℓ_j and then sum them using mlsum to form the overall log likelihood.

For instance, the probit likelihood function is

$$\ln L = \sum_{j=1}^{N} \ln \ell_j$$

$$\ln \ell_j = \begin{cases} \ln \Phi(\theta_j) & \text{if } y_j = 1 \\ \ln \Phi(-\theta_j) & \text{if } y_j = 0 \end{cases}$$

$$\theta_j = \mathbf{x}_j \mathbf{b}$$

A method d0 evaluator for probit is

```
program define myll
        version 6
        args todo b lnf
        tempvar theta lnfj
        mleval `theta' = `b'
        quietly gen double `lnfj' = ln(normprob(`theta')) if $ML_y1==1
        quietly replace `lnfj' = ln(normprob(-`theta')) if $ML_y1==0
        mlsum `lnf' = `lnfj'
end
```

In this example, we produced the individual log-likelihood values in temporary variable `` `lnfj' `` and sum `` `lnfj' `` to produce `` `lnf' ``.

Although there are other ways you could do it, we urge you to use mlsum to produce the overall sum. As a method d0 programmer, you have certain responsibilities which are easily forgotten, and which mlsum will handle for you:

1. When ml model's nopreserve option is specified, it is your responsibility to restrict the summation to the estimation subsample, the observations for which $ML_samp==1. mlsum does that for you.
2. It is your responsibility to include weights if they are specified, which are stored in $ML_w. mlsum does that for you.
3. It is your responsibility to place missing value in `` `lnf' `` if the log likelihood could not be evaluated for any relevant observation. mlsum does that for you.

It so happens that probit meets the linear-form restrictions. When the restrictions are met, writing a method d0 evaluator is easy. The method d0 evaluator differs from the method lf evaluator in that (1) you must tear apart the overall coefficient vector to produce the thetas (mleval does that) and (2) you must sum the contributions (mlsum does that). It also differs in that the equivalent method d0 evaluator will execute more slowly than the method lf evaluator.

The only reason to code a function that meets the linear-form restrictions as a method d0 evaluator is because you wish to produce a method d1 or d2 evaluator and producing a working method d0 evaluator is the best way to start.

5.1.3 Example 2: Random-effects regression

In panel data you have repeated observations on each of a collection of individual persons, firms, etc. Each observation in the data represents one of the observations on one of those individuals. For instance, if we had 5 observations on each of 20 individuals, our dataset would contain $5 \times 20 = 100$ observations in total.

The likelihood functions for panel-data estimators violate the linear-form restrictions because the observations in the dataset are not independent; likelihood values are calculated across groups of observations.

Panel-data estimators share common features and these reflect themselves in how you approach the problem:

1. Because observations are not independent, you must use ml method d0 (or d1 or d2).
2. You calculate the independent pieces of the log likelihood by sorting the data on i and then performing calculations `by` i`:`. Invariably calculations involve sums and you use constructs like

   ```
   by i: generate double ... = ... sum(...) ...
   by i: replace ... = ... sum(...) ...
   ```

   Remember that Stata's `sum()` function produces running sums; thus the completed sum is in the last observation of each group.
3. Once you have the independent pieces, you sum the contributions to obtain the overall log-likelihood value using `mlsum`. You must specify an `if` *exp* using `mlsum` in this case because you must tell `mlsum` in which observations contributions are supposed to exist.

For example, consider the random-effects regression model

$$y_{it} = \mathbf{x}_{it}\boldsymbol{\beta} + u_i + e_{it}, \quad u_i \sim N(0, \sigma_u^2), \quad e_{it} \sim N(0, \sigma_e^2)$$

The log likelihood for the ith group of observations is

$$\begin{aligned}
\ln L_i &= \ln \int_{-\infty}^{\infty} \prod_{t=1}^{T_i} f(y_{it}|u_i) f(u_i) du_i \\
&= -.5\Big[\frac{\sum_t z_{it}^2 - a_i(\sum_t z_{it})^2}{\sigma_e^2} + \ln(T_i\sigma_u^2/\sigma_e^2 + 1) + T_i \ln(2\pi\sigma_e^2)\Big] \\
T_i &= \text{number of observations in the } i\text{th group} \\
z_{it} &= y_{it} - \mathbf{x}_{it}\boldsymbol{\beta} \\
a_i &= \sigma_u^2/(T_i\sigma_u^2 + \sigma_e^2)
\end{aligned}$$

That is, we have a dataset that looks like this

	i	t	y_{it}	$\mathbf{x}_{it}$	
1.	1	1	y_{11}	$\mathbf{x}_{11}$	
2.	1	2	y_{12}	$\mathbf{x}_{12}$	
3.	1	3	y_{13}	$\mathbf{x}_{13}$	there is one log-likelihood value for these 3 observations
4.	2	1	y_{21}	$\mathbf{x}_{21}$	
5.	2	2	y_{22}	$\mathbf{x}_{22}$	
6.	2	3	y_{23}	$\mathbf{x}_{23}$	
7.	2	4	y_{24}	$\mathbf{x}_{24}$	there is one log-likelihood value for these 4 observations
8.	...				

Each group of observations, $i = 1$, $i = 2$, ..., represents, say, a person observed at various times. We have three observations for person 1 and four observations for person 2. We do not have a log-likelihood value for each observation; we have a single log-likelihood value for each group of observations and it is those values that we sum. Thus, this likelihood violates the linear-form restrictions.

A method d0 evaluator which will handle this is (we write $\theta_{1j} = \mathbf{x}_j\boldsymbol{\beta}$, $\theta_2 = \ln\sigma_u$, and $\theta_3 = \ln\sigma_e$)

```
program define myll
        version 6
        args todo b lnf
        tempvar theta1 z T S_z2 Sz_2 a
        tempname s_u s_e
        mleval `theta1´ = `b´, eq(1)
        mleval `s_u´ = `b´, eq(2) scalar
        mleval `s_e´ = `b´, eq(3) scalar
        quietly {
                gen double `z´ = $ML_y1 - `theta1´
                scalar `s_u´ = exp(`s_u´)
                scalar `s_e´ = exp(`s_e´)
                by i: gen `T´ = cond(_n==_N,_N,.)
                by i: gen double `S_z2´ = cond(_n==_N,sum(`z´^2),.)
                by i: gen double `Sz_2´ = cond(_n==_N,sum(`z´)^2,.)
                gen double `a´ = `s_u´^2 / (`T´*`s_u´^2 + `s_e´^2)
                mlsum `lnf´     = -.5 *((`S_z2´-`a´*`Sz_2´)/`s_e´^2 + /*
                                  */ ln(`T´*`s_u´^2/`s_e´^2 + 1) + /*
                                  */ `T´*ln(2*_pi*`s_e´^2) ) /*
                                  */ if `T´~=.
        }
end
```

It works. Using data from Greene (1997, 614),

```
. ml model d0 myll (lnY = lnC) /s_u /s_e
. ml max
initial:       log likelihood = -198.51759
alternative:   log likelihood = -94.908049
rescale:       log likelihood = -69.038049
rescale eq:    log likelihood = -27.524355
Iteration 0:   log likelihood = -27.524355  (not concave)
Iteration 1:   log likelihood = -20.502509  (not concave)
Iteration 2:   log likelihood = -7.9323092  (not concave)
Iteration 3:   log likelihood = -3.5463344
Iteration 4:   log likelihood =  1.8266475  (not concave)
Iteration 5:   log likelihood =  2.1716431
Iteration 6:   log likelihood =  2.6567186
Iteration 7:   log likelihood =  2.8737329  (not concave)
Iteration 8:   log likelihood =  3.0854327
Iteration 9:   log likelihood =  3.3181337
Iteration 10:  log likelihood =  3.3262447
Iteration 11:  log likelihood =  3.3262484
Iteration 12:  log likelihood =  3.3262484
```

```
                                                  Number of obs   =         24
                                                  Wald chi2(1)    =     354.24
Log likelihood =  3.3262484                       Prob > chi2     =     0.0000

------------------------------------------------------------------------------
     lnY |      Coef.   Std. Err.       z     P>|z|       [95% Conf. Interval]
---------+--------------------------------------------------------------------
eq1      |
     lnC |   1.122441   .0596367     18.821   0.000       1.005555    1.239327
   _cons |   4.713262   .2034061     23.172   0.000       4.314593     5.11193
---------+--------------------------------------------------------------------
s_u      |
   _cons |   -1.92745   .4412866     -4.368   0.000      -2.792356   -1.062545
---------+--------------------------------------------------------------------
s_e      |
   _cons |  -1.718883   .1713307    -10.033   0.000      -2.054685   -1.383081
------------------------------------------------------------------------------
```

The one thing that might escape your attention is the last line of our program:

```
mlsum `lnf' = ... if `T'~=.
```

The `if` condition at the end of the `mlsum` is of great importance. We created a log-likelihood value in the last observation of each group; variable `` `l' `` contained missing in the other observations. In performing the `mlsum`, we must specify which observations are to be summed because `mlsum` will not skip missing values. Instead, it will interpret them as meaning we tried to calculate a log-likelihood value for the observation but failed, probably because the parameter values are absurd. `mlsum` will store missing in `` `lnf' ``, flagging `ml` as to the assumed problem.

The missing values in this case, however, are intentional; we have one log-likelihood value per group of observations, so we must restrict `mlsum`'s attention to those observations by specifying an `if` *exp*. There are lots of ways we could have expressed the restriction, but we chose `` if `T'~=. `` because we had variable `` `T' `` and knew it contained nonmissing in the last observation of each group and missing everywhere else. Perhaps it would have been better in terms of understanding had we coded

```
tempvar last
by `i': generate `last' = 1 if _n==_N
mlsum `lnf' = `l' if `last'==1
```

but it seemed silly to waste the computer's time when `` if `T'~=. `` would do just as well.

5.2 Aside: Stata's scalars

We are about to have a long aside because, to use method d0 effectively—and methods d1 and d2—you need to use Stata's scalars. If, in your likelihood evaluator, you need to make intermediate scalar calculations, do not code

```
local eta = ...
generate double ... = ... `eta' ...
```

Code instead

```
tempname eta
scalar `eta' = ...
generate double ... = ... `eta' ...
```

You do this for the same reason that the intermediate variables you `generate` are made `doubles`; scalars are more accurate than macros. Macros provide about 12 digits of accuracy, sometimes more but never less. Scalars are full double-precision binary numbers, meaning they provide about 16.5 digits of accuracy in all cases.

A Stata scalar is just that—it contains a single number (which could be a missing value). The **scalar** command handles scalar definition and you use scalars in expressions just as you would any variable:

```
. scalar x = 3
. scalar y = 1/sqrt(2)
. display x
3
. display y
.70710678
. scalar z = 1/ln(x-3)
. display z
.
```

We have been sloppy in the above, although there is no reason you would know that. Here is how the example would have better read:

```
. scalar x = 3
. scalar y = 1/sqrt(2)
. display scalar(x)                  ← We type scalar(x), not x
3
. display scalar(y)                  ← We type scalar(y), not y
.70710678
. scalar z = 1/ln(scalar(x)-3)       ← We type scalar(x), not x
. display scalar(z)                  ← We type scalar(z), not z
.
```

The **scalar()** function says "use the value of the named scalar here" and, as you have seen, we can omit it. We should not, however, unless we are certain there is no variable in the dataset by the same name. Pretend we had a variable named **x** and a scalar named **x**:

```
. clear
. input x                            ← create the variable named x
             x
  1. 1
  2. 2
  3. 3
  4. end
. scalar x = 57                      ← create the scalar named x
. gen y = x+1                        ← what is in y?
```

Notice that "**gen y = x+1**" did not result in an error, nor did defining a scalar named **x** when a variable named **x** already existed; neither would it have been an error to do things the other way around and create a variable named **x** when a scalar named **x** already existed. This, however, leads to an interesting question. To which are we referring when we type **x** in "**gen y = x+1**". Is it the data variable, in which case **y** would contain 2, 3 and 4, or is it the scalar, in which case **y** would contain 58, 58, and 58?

When Stata has a variable and scalar of the same name, it chooses the variable in preference to the scalar:

```
. list
                                          We previously typed gen y = x + 1.
               x          y
  1.           1          2               There is a scalar named x and it
  2.           2          3               contains 57. Nevertheless, we
  3.           3          4               obtained y = VARIABLE x plus 1.
```

If we really wanted Stata to use the scalar called `x`, we must tell it:

```
. gen z = scalar(x) + 1
. list
              x          y          z
  1.          1          2         58
  2.          2          3         58
  3.          3          4         58
```

The same logic applies to all Stata commands, including **scalar** itself:

```
. scalar new = x
. display new
1
```

Why 1 and not 57? Because "**scalar new** = x" was interpreted as

```
. scalar new = /*VARIABLE*/ x
```

and that in turn was interpreted as

```
. scalar new = /*VARIABLE*/ x /*IN THE FIRST OBSERVATION*/
```

Had there instead been no variable called `x`, "**scalar new** = x" would have been interpreted as

```
. scalar new = /*SCALAR*/ x
```

If there were no variable and no scalar named `x`, then "**scalar new** = x" would have resulted in the error message, "`x` not found", r(111).

Stata's preference for variables over scalars is even stronger than we have indicated. Here is another example:

```
. clear
. input mpg weight
             mpg     weight
  1. 22 2100
  2. 23 1900
  3. 12 3800
  4. end
. scalar m = 9.5               ← create a scalar called x
. scalar z = m                 ← set z equal to m
. display z
22                             ← scalar z contains 22 and not 9.5 (!)
```

Scalar `z` contains 22 and not 9.5 because `m` was taken as an abbreviation for variable `mpg`! Thus, it appears that about the only safe way we can use scalars is to reference them inside the **scalar()** function:

```
. scalar z = scalar(m)
. display z
9.5
```

In fact, we seldom do that. Instead, we write our programs using the **tempname** statement:

```
tempname x y
scalar `x´ = (some complicated function)
scalar `y´ = ... `x´ ...
```

The last line could have read

```
scalar `y´ = ... scalar(`x´) ...
```

but the **scalar()** is not necessary. **tempname** will create a unique name for the scalar and we can be certain that there is no variable by the same name, abbreviated or not.

5.3 Example 3: The Cox proportional hazards model

The Cox proportional hazards model has a reputation of being difficult to calculate, to say it mildly.

Let $\mathbf{x}_j$ be a (fixed) row vector of covariates for a person. The hazard of failure for the person is assumed to be

$$h_j(t) = h_0(t)\exp(\mathbf{x}_j\mathbf{b})$$

The Cox proportional hazards model provides estimates of $\mathbf{b}$, leaving $h_0(t)$ unestimated. Given a dataset of N individuals in which each individual is observed from time 0 to t_j, at which time the individual either is observed to fail or is censored, the likelihood function is

$$\ln L = \sum_{i=1}^{N}\Big[\sum_{k\in D_i}\mathbf{x}_k\mathbf{b} \quad - \quad d_i \ln\Big(\sum_{j\in R_i}\exp[\mathbf{x}_j\mathbf{b}]\Big)\Big]$$

where i indexes the ordered failure times $t_{(i)}$ $(i = 1,\dots,N)$, d_i is the number who fail at that time, D_i is the set of observations that fail at that time, and R_i is the set of observations that are at risk at time $t_{(i)}$. This likelihood has the property that

1. there are likelihood values only when deaths occur and
2. the likelihood is a function of all the observations who have not died yet.

As such, it violates the lf assumptions. Moreover, it is not an easy likelihood function to calculate in any language.

To establish some easier-to-type notation, let us rewrite the likelihood function as

$$\ln L = \sum_{i=1}^{N}\big[A_i - d_i\ln(B_i)\big]$$

where

$$A_i = \sum_{k\in D_i}\mathbf{x}_k\mathbf{b}$$

$$B_i = \sum_{j\in R_i}\exp(\mathbf{x}_j\mathbf{b})$$

Actually, we rewrote this likelihood function for another reason, too. When facing a difficult problem, it helps to break it into more approachable pieces. Mechanically, the method is to remove some complicated term from the likelihood function, give it a letter, substitute the letter in the original formula, and write the definition of the letter on the side. Once you have done that, you think about how to program each of the individual ingredients. As things stand right now, if we knew A_i, d_i, and B_i, calculating $\ln L$ would be easy: $\ln L = \sum A_i - d_i\ln(B_i)$. So now we think about how to obtain A_i, d_i, and B_i.

To fix ideas, let's pretend we had the following data:

```
id   t     d      x1  ... xk
------------------------------
 1   1     1      ...
 2   3     0      ...
 3   5     1      ...
 4   5     1      ...
 5   5     0      ...
 6   6     1      ...
```

`t` is time of death or censoring
`d` is 1 if death, 0 if censoring

A person is in the risk pool
until they die or are censored.

In terms of calculating $B_i = \sum_{\text{those still alive}} \exp(\mathbf{x}_j \mathbf{b})$, let us think about the pool of still-alive persons:

```
id   t    d     x1  ... xk
------------------------------
 1   1    1     ...                ← 1st death, everybody alive
 2   3    0     ...
 3   5    1     ...                ← 2nd and 3rd deaths...
 4   5    1     ...                ← ids 1 and 2 are out of pool
 5   5    0     ...
 6   6    1     ...                ← 4th death, all ids above are out
```

If you read the above table from bottom to top, the calculation of who is in the pool becomes easy: At the bottom, nobody is in the pool, then person 6 is added to the pool, then person 5, and so on. So we could reverse the data and calculate as follows:

```
id   t    d     x1  ... xk   B
-----------------------------------   pool is empty, B = 0
 6   6    1     ...                   add to pool, B = B + exp(X_6*b)
 5   5    0     ...                   add to pool, B = B + exp(X_5*b)
 4   5    1     ...                   add to pool, B = B + exp(X_4*b)
 3   5    1     ...                   add to pool, B = B + exp(X_3*b)
 2   3    0     ...                   add to pool, B = B + exp(X_2*b)
 1   1    1     ...                   add to pool, B = B + exp(X_1*b)
```

Statawise, this is easy to do:

```
gen double negt = -t
sort negt d
gen double B = sum(exp(`xb'))
```

where `` `xb' `` is just the evaluation of $\mathbf{x}_j \mathbf{b}$:

```
args todo b lnf
tempvar xb
mleval `xb' = `b'
```

In previous examples, we have referred to this as `` `theta' `` but we wanted something shorter and more meaningful. There is no rule that says the evaluation of the equation has to be named `` `theta' ``.

Calculating $A_i = \sum_{\text{those who die at a particular time}} \mathbf{x}_k \mathbf{b}$ is no more difficult:

```
by negt d: gen double A = cond(_n==_N, sum(`xb'), .) if d==1
```

We need one more ingredient, d_i, the number who die at each time:

```
by negt d: gen double sumd = cond(_n==_N, sum(d), .)
```

With `A`, `B`, and `sumd`, calculating the log-likelihood value is easy:

```
gen double L = A - sumd*ln(B)
by negt d: gen byte last = (_n==_N & d==1)
mlsum `lnf' = L if last
```

Note the `if last` on the `mlsum` statement; it is of vital importance. By default, `mlsum` assumes that every observation should contribute to the likelihood value. If there are any missing values among the contributions, `mlsum` assumes that means the likelihood cannot be calculated at the given value of $\mathbf{b}$. It fills in the overall log likelihood `` `lnf' `` with missing and then later, when ml sees that result, it takes the appropriate action. We have lots of missing values in `L`, and purposefully, because we have contributions to the likelihood only when there is a failure. Thus, we need to mark the observations that should contribute to the likelihood and tell `mleval` about it. We were, we admit, tempted to code

```
gen double L = A - sumd*ln(B)
mlsum `lnf' = L if L~=.
```

because then we would sum the contributions that do exist. That would have been a bad idea. Perhaps, given **b**, one of the contributions that should exist could not be calculated, meaning that the overall likelihood really could not be calculated at **b**. We want to be sure that, should that event arise, `mlsum` sets the overall $\ln L$ in `` `lnf' `` to missing so that ml takes the appropriate action. Therefore, we carefully marked which values should be filled in and specified that on the `mlsum` statement.

When programming something this complicated, it is a good idea to assemble a summary sheet from which you then write your code.

5.3.1 Problem summary sheet

To be estimated is

$$\text{relative hazard} = \exp(x_{1j}b_1 + x_{2j}b_2 + \cdots + x_{kj}b_k)$$

Note that this model has no intercept. We will type the `ml model` statement

```
. ml model d0 coxlf (timevar failvar = x1 x2 ... xk, nocons)
```

The likelihood-evaluation code will read, in outline,

```
args todo b lnf
mleval xb = `b'

gen double negt = -t
sort negt d

gen double B = sum(exp(xb))

by negt d: gen double A = cond(_n==_N, sum(xb), .) if d==1
by negt d: gen sumd = cond(_n==_N, sum(d), .)

gen double L = A - sumd*ln(B)
by negt d: gen byte last = (_n==_N & d==1)
mlsum `lnf' = L if last
```

Temporary variables in the above are

```
tempvar xb B A sumd L negt last
```

and obviously we need to go back and insert single quotes when we code this for real.

5.3.2 Do-file to perform estimation

Based on the summary sheet, we wrote the following do-file to estimate the particular model

$$\text{relative hazard} = \exp(b_1\texttt{drug2} + b_2\texttt{drug3} + b_3\texttt{age})$$

using the `cancer.dta` that comes with Stata. The dataset has variables

`studytim`	Months to death or censoring, t in our notation
`died`	1 if patient died, d in our notation
`drug`	drug type 1, 2, or 3
`age`	patient's age at start of experiment

Here is our do-file:

top of coxd0.do

```
clear
program drop _all

program define myll
        version 6
        args todo b lnf
        local t "$ML_y1"                    /* make t a synonym for $ML_y1 */
        local d "$ML_y2"                    /* make d a synonym for $ML_y2 */
        tempvar xb B A sumd L negt last
        mleval `xb´ = `b´
        gen double `negt´ = -`t´
        sort `negt´ `d´
        quietly {
                gen double `B´ = sum(exp(`xb´))
                by `negt´ `d´: gen double `A´ = cond(_n==_N, sum(`xb´), .) if `d´==1
                by `negt´ `d´: gen `sumd´ = cond(_n==_N, sum(`d´), .)
                gen double `L´ = `A´ - `sumd´*ln(`B´)
                by `negt´ `d´: gen byte `last´ = (_n==_N & `d´==1)
        }
        mlsum `lnf´ = `L´ if `last´
end

use cancer, clear
tab drug, gen(dr)      /* make dummies dr1, dr2, dr3 */

ml model d0 myll (studytim died = dr2 dr3 age, nocons)
ml max
```

end of cox0.do

The result of running it is:

```
. do coxd0
  (output omitted )

. ml model d0 myll (studytim died = dr2 dr3 age, nocons)

. ml max

initial:       log likelihood = -99.911448
alternative:   log likelihood = -99.606964
rescale:       log likelihood = -99.101334
Iteration 0:   log likelihood = -99.101334
Iteration 1:   log likelihood = -82.649192
Iteration 2:   log likelihood = -81.733652
Iteration 3:   log likelihood = -81.652743
Iteration 4:   log likelihood = -81.652567
Iteration 5:   log likelihood = -81.652567

                                                  Number of obs   =         48
                                                  Wald chi2(3)    =      28.33
Log likelihood = -81.652567                       Prob > chi2     =     0.0000

------------------------------------------------------------------------------
         |      Coef.   Std. Err.       z     P>|z|       [95% Conf. Interval]
---------+--------------------------------------------------------------------
     dr2 |   -1.71156   .4943639     -3.462   0.001      -2.680495   -.7426242
     dr3 |  -2.956384   .6557432     -4.508   0.000      -4.241617   -1.671151
     age |     .11184   .0365789      3.058   0.002       .0401467    .1835333
------------------------------------------------------------------------------
```

Those are the correct answers.

Our program is not as efficient as it might be. Every time our likelihood evaluator is called it resorts the data. For your information, `ml` called our evaluator 116 times in the example above—which we determined using `ml count` (see **A. Syntax diagrams**). Clearly, sorting the data just once would save considerable execution time.

One easy solution to this would be to change our do-file so that the we specified the negative survival time on the `ml model` statement and we sorted the data once prior to execution:

top of coxd0s.do

```
clear
program drop _all
program define myll
        version 6
        args todo b lnf
        local negt "$ML_y1"              /* make negt a synonym for $ML_y1 */
        local d "$ML_y2"                 /* make d a synonym for $ML_y2 */
        tempvar xb B A sumd L last
        mleval `xb' = `b'
        quietly {
                gen double `B' = sum(exp(`xb'))
                by `negt' `d': gen double `A' = cond(_n==_N, sum(`xb'), .) if `d'==1
                by `negt' `d': gen `sumd' = cond(_n==_N, sum(`d'), .)
                gen double `L' = `A' - `sumd'*ln(`B')
                by `negt' `d': gen byte `last' = (_n==_N & `d'==1)
        }
        mlsum `lnf' = `L' if `last'
end
use cancer, clear
tab drug, gen(dr)       /* make dummies dr1, dr2, dr3 */
gen negtime = -studytim
sort negtime died
ml model d0 myll (negtime died = dr2 dr3 age, nocons)
ml max
```

end of coxd0s.do

We are unsure whether to encourage this. The do-file is only a little faster because, in the original version, Stata did not actually sort the data 116 times. `sort` is smart and, when told to sort, it looks at the sort markers to determine if the dataset is already sorted. Thus, the original code sorted the data once and made `sort` verify that it was sorted 115 times.

Still, there are situations where there is a considerable gain to making calculations once and only once. You can do that, but be aware that you make your do-file a little more dangerous to use.

❑ Technical Note

If we were using our evaluator as a subroutine in writing a new estimation command, we could gain the efficiency and be safe about it. We would be calling `ml` from inside another ado-file and so have complete control. We could create `` `negt' `` and sort on it before the `ml model` statement and then use `` `negt' `` in the `ml model` statement. The user could not forget to sort the data, or mistakenly use time rather than negative time, because we would be the ones doing that behind the scenes.

❑

5.4 References

Greene, W. H. 1997. *Econometric Analysis.* 3d ed. Upper Saddle River, NJ: Prentice–Hall.

6 Method d1

Contents

6.1 Basic syntax

Method d1 may be used with any likelihood function,

$$
\begin{aligned}
\ln L &= \ln L\big((\theta_{1j}, \theta_{2j}, \ldots, \theta_{Ej}; y_{1j}, y_{2j}, \ldots, y_{Dj}) : j = 1, \ldots, N\big) \\
\theta_{1j} &= \mathbf{x}_{1j}\mathbf{b}_1 \\
\theta_{2j} &= \mathbf{x}_{2j}\mathbf{b}_2 \\
&\;\vdots \\
\theta_{Ej} &= \mathbf{x}_{Ej}\mathbf{b}_E
\end{aligned}
$$

where j indexes observations.

The method d1 evaluator is to calculate $\ln L$, the overall log likelihood and its first derivatives. The outline for the program is

```
program define myprog
        version 6
        args todo b lnf g
        tempvar theta1 theta2 ...
        mleval `theta1´ = `b´
        mleval `theta2´ = `b´, eq(2) /* if there is a θ2 */
        ...
        /* if you need to create any intermediate results: */
        tempvar tmp1 tmp2 ...
        gen double `tmp1´ = ...
        ...
        mlsum `lnf´ = ...
        if `todo´==0 | `lnf´==. { exit }
        tempname d1 d2 ...
        mlvecsum `lnf´ `d1´ = formula for ∂ ln ℓj/∂θ1j, eq(1)
        mlvecsum `lnf´ `d2´ = formula for ∂ ln ℓj/∂θ2j, eq(2)
        ...
        matrix `g´ = (`d1´,`d2´,...)
end
```

1. You are passed the parameter vector. Use `mleval` to obtain the thetas from it; see **4.2.3 Using mleval**.

2. Access the dependent variables (if any) by referring to `$ML_y1`, `$ML_y2`, ... Use `$ML_y1`, `$ML_y2`, ..., just as you would any existing variable in the data.
3. The resulting log likelihood is to be saved in scalar `` `lnf' ``. Use `mlsum` to produce it; see **4.2.6 Using mlsum**.
4. The resulting gradient vector is to be saved in vector `` `g' ``. Using `mlvecsum` to produce it is appropriate only when the likelihood function meets the linear-form restrictions; see **4.2.8 Using mlvecsum**.
5. It is of great importance that, if you create any temporary variables for intermediate results, you create them as `doubles`: Type "`gen double` *name* `=` ..."; do not omit the word `double`.
6. The estimation subsample is `$ML_samp==1`. You may safely ignore this if you do not specify `ml model`'s `nopreserve` option. When your program is called, only relevant data will be in memory. If you do specify `nopreserve`, understand that `mleval`, `mlsum`, and `mlvecsum` automatically restrict themselves to the estimation subsample; it is not necessary to code `if $ML_samp==1` on these commands. You merely need to restrict other commands to the `$ML_samp==1` subsample.
7. The weights are stored in variable `$ML_w`, which contains 1 in every observation if no weights are specified. If you use `mlsum` to produce the log likelihood and `mlvecsum` to produce the gradient vector, you may ignore this because these commands automatically include the weights.
8. While method d1 evaluators can be used with `fweights`, `aweights`, and `iweights`, you must code your program a special way if `pweights` are to be specified. You must also do this if `ml model`'s `robust` or `cluster()` options are to be specified. Change the `args` statement at the top of the program from

```
args todo b lnf g
```

to

```
args todo b lnf g negH g1 g2 ...
```

Change the part of the outline that reads

```
tempname d1 d2 ...
mlvecsum `lnf' `d1' = formula for ∂ ln ℓj/∂θ1j , eq(1)
mlvecsum `lnf' `d2' = formula for ∂ ln ℓj/∂θ2j , eq(2)
...
```

to read

```
quietly replace `g1' = formula for ∂ ln ℓj/∂θ1j
quietly replace `g2' = formula for ∂ ln ℓj/∂θ2j
...
tempname d1 d2 ...
mlvecsum `lnf' `d1' = `g1', eq(1)
mlvecsum `lnf' `d2' = `g2', eq(2)
...
end
```

The above will work if the likelihood function meets the linear-form restrictions. Otherwise, `pweights`, `robust`, and `cluster()` will not work.

You specify that you are using method d1 on the `ml model` statement,

```
. ml model d1 myprog ...
```

which you issue after you have defined the likelihood-evaluation program.

6.1.1 Using method d1

The method d1 evaluator simply picks up where the method d0 evaluator left off. Taking out the code in common, here is how they differ:

```
program define myprog
        version 6
        args todo b lnf g                          /* g is added */
        ...
        if `todo'==0 | `lnf'==. { exit }           /* from here down is new */
        tempname d1 d2 ...
        mlvecsum `lnf' `d1' = formula for ∂ln ℓj/∂θ1j, eq(1)
        mlvecsum `lnf' `d2' = formula for ∂ln ℓj/∂θ2j, eq(2)
        ...
        matrix `g' = (`d1',`d2',...)
end
```

The gradient vector you are to define is

$$\mathbf{g} = \frac{\partial \ln L}{\partial \mathbf{b}} = \left(\frac{\partial \ln L}{\partial b_1}, \frac{\partial \ln L}{\partial b_2}, \ldots, \frac{\partial \ln L}{\partial b_k} \right)$$

`ml` provides the coefficient vector as a $1 \times k$ row vector and expects the gradient vector to be returned as a $1 \times k$ row vector.

If your likelihood function meets the linear-form restrictions, writing a method d1 evaluator is easy because `mlvecsum` will create the vector for you from the derivatives with respect to the thetas.

6.1.2 Example: Probit

We begin with probit because we want to start with a single-equation (single θ) system. The probit log-likelihood function is

$$\ln L = \sum_{j=1}^{N} \ln \ell_j$$

$$\ln \ell_j = \begin{cases} \ln \Phi(\theta_j) & \text{if } y_j = 1 \\ \ln \Phi(-\theta_j) & \text{if } y_j = 0 \end{cases}$$

$$\theta_j = \mathbf{x}_j \mathbf{b}$$

A working method d0 evaluator for probit is

```
program define myll
        version 6
        args todo b lnf
        tempvar theta lnfj
        mleval `theta' = `b'
        quietly gen double `lnfj' = ln(normprob(`theta')) if $ML_y1==1
        quietly replace `lnfj' = ln(normprob(-`theta')) if $ML_y1==0
        mlsum `lnf' = `lnfj'
end
```

When a likelihood meets the linear-form restrictions, it is not necessary to calculate the vector $d\ln L/d\mathbf{b}$. Instead one can calculate $\partial \ln \ell_j / \partial \theta_j$ and use `mleval` to construct the vector from that. The derivative of the probit log likelihood with respect to θ_j is

$$\frac{\partial \ln \ell_j}{\partial \theta_j} = \begin{cases} \phi(\theta_j)/\Phi(\theta_j) & \text{if } y_j = 1 \\ -\phi(\theta_j)/\Phi(-\theta_j) & \text{if } y_j = 0 \end{cases}$$

where $\phi()$ is the unit-normal density. Thus, a method d1 evaluator for probit is

```
program define myll
        version 6
        args todo b lnf g
        tempvar theta lnfj
        mleval `theta' = `b'
        quietly gen double `lnfj' = ln(normprob(`theta')) if $ML_y1==1
        quietly replace `lnfj' = ln(normprob(-`theta')) if $ML_y1==0
        mlsum `lnf' = `lnfj'
        if `todo'==0 | `lnf'==. { exit }
        tempvar gj
        quietly gen double `gj' = normd(`theta')/normprob(`theta') /*
                                */ if $ML_y1==1
        quietly replace `gj' = -normd(`theta')/normprob(-`theta') /*
                                */ if $ML_y1==0
        mlvecsum `lnf' `g' = `gj'
end
```

```
. ml model d1 myll (for=mpg weight)
. ml maximize
initial:       log likelihood = -51.292891
alternative:   log likelihood = -45.055272
rescale:       log likelihood = -45.055272
Iteration 0:   log likelihood = -45.055272
Iteration 1:   log likelihood = -27.904623
Iteration 2:   log likelihood = -26.858226
Iteration 3:   log likelihood = -26.844198
Iteration 4:   log likelihood = -26.844189
Iteration 5:   log likelihood = -26.844189
                                                  Number of obs   =         74
                                                  Wald chi2(2)    =      20.75
Log likelihood = -26.844189                       Prob > chi2     =     0.0000

------------------------------------------------------------------------------
 foreign |      Coef.   Std. Err.       z     P>|z|       [95% Conf. Interval]
---------+--------------------------------------------------------------------
     mpg |  -.1039503   .0515689     -2.016   0.044      -.2050235   -.0028772
  weight |  -.0023355   .0005661     -4.126   0.000       -.003445   -.0012261
   _cons |   8.275464   2.554142      3.240   0.001       3.269437    13.28149
------------------------------------------------------------------------------
```

6.1.3 Example: Weibull regression

Weibull is an example of a two-equation system. The important programming difference is that this time we will have to issue two `mlvecsum` commands and then use `matrix` *name*=... to put the pieces together.

The likelihood function for Weibull regression in the log-time metric is

$$\begin{aligned}\ln \ell_j &= (t_j e^{\theta_{1j}})^{\exp(\theta_{2j})} + d_j\big(\theta_{2j} - \theta_{1j} + (e^{\theta_{2j}} - 1)(\ln t_j - \theta_{1j})\big)\\ \theta_{1j} &= \mathbf{x}_j\mathbf{b}\\ \theta_{2j} &= s\end{aligned}$$

where t_j is the time of death or censoring and d_j is 1 if death and 0 if censored. The parameters to be estimated are $\mathbf{b}$ and s. It will be easier to write the likelihood evaluation programs if we rewrite the likelihood function with some intermediate variables. The log-likelihood function can be more compactly written

$$\begin{aligned}
\ln \ell_j &= -M_j + d_j\big(\theta_{2j} - \theta_{1j} + (p_j - 1)R_j\big) \\
p_j &= e^{\theta_{2j}} \\
M_j &= (t_j e^{-\theta_{1j}})^{p_j} \\
R_j &= \ln t_j - \theta_{1j}
\end{aligned}$$

It is important to understand that we would never attempt to produce a method d1 evaluator without writing a method d0 evaluator first. A working method d0 evaluator for this likelihood is

```
program define weib0
        version 6
        args todo b lnf
        tempvar theta1 theta2
        mleval `theta1' = `b', eq(1)
        mleval `theta2' = `b', eq(2)
        local t "$ML_y1"            /* this is just for readability */
        local d "$ML_y2"
        tempvar p M R
        quietly gen double `p' = exp(`theta2')
        quietly gen double `M' = (`t'*exp(-`theta1'))^`p'
        quietly gen double `R' = ln(`t')-`theta1'
        mlsum `lnf' = -`M' + `d'*(`theta2'-`theta1' + (`p'-1)*`R')
end
```

The first derivatives of the Weibull log-likelihood function with respect to the thetas are

$$\begin{aligned}
\frac{\partial \ln \ell_j}{\partial \theta_{1j}} &= p_j(M_j - d_j) \\
\frac{\partial \ln \ell_j}{\partial \theta_{2j}} &= d_j - R_j p_j(M_j - d_j)
\end{aligned}$$

Thus, to convert our working d0 evaluator into a d1 evaluator, it is just a matter of calculating these two formulas:

```
program define weib1
        version 6
        args todo b lnf g                                          /* g is new */
        tempvar theta1 theta2
        mleval `theta1' = `b', eq(1)
        mleval `theta2' = `b', eq(2)
        local t "$ML_y1"
        local d "$ML_y2"
        tempvar p M R
        quietly gen double `p' = exp(`theta2')
        quietly gen double `M' = (`t'*exp(-`theta1'))^`p'
        quietly gen double `R' = ln(`t')-`theta1'
        mlsum `lnf' = -`M' + `d'*(`theta2'-`theta1' + (`p'-1)*`R')
        if `todo'==0 | `lnf'==. { exit }                          /* <-- new */
        tempname d1 d2                                             /* <-- new */
        mlvecsum `lnf' `d1' = `p'*(`M'-`d'), eq(1)                 /* <-- new */
        mlvecsum `lnf' `d2' = `d' - `R'*`p'*(`M'-`d'), eq(2)      /* <-- new */
        matrix `g' = (`d1',`d2')                                   /* <-- new */
end
```

This routine produces correct answers:

```
. use cancer, clear
(Patient Survival in Drug Trial)
. quietly tabulate drug, gen(drug)
. ml model d1 weib1 (studytim died = drug2 drug3 age) /s
. ml maximize
initial:       log likelihood =       -744
alternative:   log likelihood = -356.14276
rescale:       log likelihood = -200.80201
rescale eq:    log likelihood = -136.69232
Iteration 0:   log likelihood = -136.69232  (not concave)
Iteration 1:   log likelihood = -124.13044
Iteration 2:   log likelihood =  -113.9905
Iteration 3:   log likelihood = -110.30732
Iteration 4:   log likelihood = -110.26748
Iteration 5:   log likelihood = -110.26736
Iteration 6:   log likelihood = -110.26736
                                                  Number of obs   =         48
                                                  Wald chi2(3)    =      35.25
Log likelihood = -110.26736                       Prob > chi2     =     0.0000
------------------------------------------------------------------------------
             |      Coef.   Std. Err.      z    P>|z|     [95% Conf. Interval]
-------------+----------------------------------------------------------------
eq1          |
       drug2 |   1.012966   .2903917     3.488   0.000      .4438087   1.582123
       drug3 |    1.45917   .2821195     5.172   0.000      .9062261   2.012114
         age |  -.0671728   .0205687    -3.266   0.001     -.1074868  -.0268587
       _cons |   6.060723   1.152845     5.257   0.000      3.801189   8.320258
-------------+----------------------------------------------------------------
s            |
       _cons |   .5573333   .1402153     3.975   0.000      .2825163   .8321504
------------------------------------------------------------------------------
```

If we compared results with Stata's `weibull` or `streg, dist(weibull)` commands, we would note one difference: the reported value of the log likelihood is different. In the formulas above we did not include certain constants which, while having no formal role in the Weibull model, are usually included because they normalize away the units in which time are recorded. This difference is of no importance and it would just complicate our program were we to include them.

6.2 Method d1debug

We have proceeded too quickly. In real life the probit and Weibull method d1 evaluators would not have worked the first time we tried. Without a doubt we would have made an error in producing the gradient vector, either substantively because we took the derivative incorrectly, or mechanically because we coded mistakenly.

The recipe for success is to start with a working method d0 evaluator, make the changes necessary to convert it to method d1, but then specify the method d1debug on the `ml model` statement. Method d1debug is just like method d1 except that it checks your derivatives against numerically calculated ones.

For instance, here is what happens when we test the working `weib1` using method d1debug:

```
. ml model d1debug weib1 (studytim died = drug2 drug3 age) /s
. ml maximize
initial:       log likelihood =       -744
alternative:   log likelihood = -356.14276
rescale:       log likelihood = -200.80201
```

```
rescale eq:     log likelihood = -136.69232

d1debug:  Begin derivative-comparison report ----------------------------------
d1debug:  mreldif(gradient vector) =  .0021855
d1debug:  End derivative-comparison report ------------------------------------
Iteration 0:    log likelihood = -136.69232  (not concave)

d1debug:  Begin derivative-comparison report ----------------------------------
d1debug:  mreldif(gradient vector) =  7.17e-10
d1debug:  End derivative-comparison report ------------------------------------
Iteration 1:    log likelihood = -124.10384

d1debug:  Begin derivative-comparison report ----------------------------------
d1debug:  mreldif(gradient vector) =  7.70e-10
d1debug:  End derivative-comparison report ------------------------------------
Iteration 2:    log likelihood = -112.12286

d1debug:  Begin derivative-comparison report ----------------------------------
d1debug:  mreldif(gradient vector) =  1.41e-09
d1debug:  End derivative-comparison report ------------------------------------
Iteration 3:    log likelihood = -110.29748

d1debug:  Begin derivative-comparison report ----------------------------------
d1debug:  mreldif(gradient vector) =  .0000264
d1debug:  End derivative-comparison report ------------------------------------
Iteration 4:    log likelihood = -110.26738

d1debug:  Begin derivative-comparison report ----------------------------------
d1debug:  mreldif(gradient vector) =    .00007
d1debug:  End derivative-comparison report ------------------------------------
Iteration 5:    log likelihood = -110.26736

d1debug:  Begin derivative-comparison report ----------------------------------
d1debug:  mreldif(gradient vector) =  .0000701
d1debug:  End derivative-comparison report ------------------------------------
Iteration 6:    log likelihood = -110.26736

                                                  Number of obs   =         48
                                                  Wald chi2(3)    =      35.25
Log likelihood = -110.26736                       Prob > chi2     =     0.0000

------------------------------------------------------------------------------
         |      Coef.   Std. Err.       z     P>|z|       [95% Conf. Interval]
---------+--------------------------------------------------------------------
eq1      |
   drug2 |   1.012966   .2903919      3.488   0.000        .4438084    1.582124
   drug3 |    1.45917   .2821196      5.172   0.000        .9062261    2.012115
     age |  -.0671728    .020569     -3.266   0.001       -.1074872   -.0268584
   _cons |   6.060723   1.152856      5.257   0.000        3.801166    8.320281
---------+--------------------------------------------------------------------
s        |
   _cons |   .5573333   .1402154      3.975   0.000        .2825161    .8321504
------------------------------------------------------------------------------
```

The difference between methods d1 and d1debug is that d1debug provides a derivative comparison report interspersed in the iteration log. For instance, between iterations 1 and 2 we see the following:

```
Iteration 1:    log likelihood = -124.10384

d1debug:  Begin derivative-comparison report ----------------------------------
d1debug:  mreldif(gradient vector) =  7.70e-10
d1debug:  End derivative-comparison report ------------------------------------
Iteration 2:    log likelihood = -112.12286
```

Whenever `ml` needs first derivatives, method d1debug calls your likelihood evaluator just as method d1 would, but method d1debug also calculates the derivatives numerically. It then displays a comparison of the derivatives calculated both ways and it returns to `ml` the *numerically calculated derivatives*. Thus, d1debug proceeds to a correct answer even if your derivative calculation is wrong.

`mreldif()` in the derivative-comparison report stands for maximum relative difference; this would be easier to understand if we could see the gradient vector. Method d1debug will show us the vectors as well as the `mreldif()` if we specify `ml maximize`'s `gradient` option:

```
. ml model d1debug weib1 (studytim died = drug2 drug3 age) /s

. ml maximize, gradient

initial:       log likelihood =      -744
alternative:   log likelihood = -356.14276
rescale:       log likelihood = -200.80201
rescale eq:    log likelihood = -136.69232
------------------------------------------------------------------------------
Iteration 0:

d1debug:  Begin derivative-comparison report ---------------------------------
d1debug:  mreldif(gradient vector) =  .0021855

d1debug:  weib1-calculated gradient:
             eq1:        eq1:        eq1:        eq1:           s:
           drug2       drug3         age       _cons        _cons
r1  -.7819348   1.243378   -632.982  -10.67705   2.580612

d1debug:  numerically calculated gradient (used for stepping):
             eq1:        eq1:        eq1:        eq1:           s:
           drug2       drug3         age       _cons        _cons
r1  -.7782977   1.245565  -632.9787  -10.67703   2.572804

d1debug:  relative difference:
           eq1:       eq1:       eq1:       eq1:         s:
         drug2      drug3        age      _cons      _cons
r1  .0020452  .0009738  5.14e-06  1.87e-06  .0021855

d1debug:  End derivative-comparison report -----------------------------------
                                               log likelihood = -136.69232
                                                             (not concave)
------------------------------------------------------------------------------
Iteration 1:

d1debug:  Begin derivative-comparison report ---------------------------------
d1debug:  mreldif(gradient vector) =  7.17e-10

d1debug:  weib1-calculated gradient:
            eq1:       eq1:       eq1:       eq1:         s:
          drug2      drug3        age      _cons      _cons
r1  12.07714    11.9357   1356.576  25.82583  7.391241

d1debug:  numerically calculated gradient (used for stepping):
            eq1:       eq1:       eq1:       eq1:         s:
          drug2      drug3        age      _cons      _cons
r1  12.07714    11.9357   1356.576  25.82583  7.391241

d1debug:  relative difference:
            eq1:       eq1:       eq1:       eq1:         s:
          drug2      drug3        age      _cons      _cons
r1  2.12e-10  1.37e-11  7.17e-10  7.17e-10  1.59e-10

d1debug:  End derivative-comparison report -----------------------------------
                                               log likelihood = -124.10384
------------------------------------------------------------------------------
Iteration 2:

d1debug:  Begin derivative-comparison report ---------------------------------
d1debug:  mreldif(gradient vector) =  7.70e-10

d1debug:  weib1-calculated gradient:
             eq1:        eq1:        eq1:        eq1:           s:
           drug2       drug3         age       _cons        _cons
r1   .6561854   1.037603   82.27463   2.673944  -11.22783
```

```
d1debug:  numerically calculated gradient (used for stepping):
            eq1:        eq1:        eq1:        eq1:          s:
          drug2       drug3         age       _cons       _cons
r1    .6561854    1.037603    82.27463    2.673944   -11.22783
d1debug:  relative difference:
            eq1:        eq1:        eq1:        eq1:          s:
          drug2       drug3         age       _cons       _cons
r1    6.05e-10    7.70e-10    7.93e-11    3.00e-10    7.00e-10
d1debug:  End derivative-comparison report ----------------------------------
```

(output omitted)

```
                                                  Number of obs   =         48
                                                  Wald chi2(3)    =      35.25
Log likelihood = -110.26736                       Prob > chi2     =     0.0000

------------------------------------------------------------------------------
         |      Coef.   Std. Err.       z     P>|z|       [95% Conf. Interval]
---------+--------------------------------------------------------------------
eq1      |
   drug2 |   1.012966   .2903919     3.488   0.000       .4438084    1.582124
   drug3 |    1.45917   .2821196     5.172   0.000       .9062261    2.012115
     age |  -.0671728    .020569    -3.266   0.001      -.1074872   -.0268584
   _cons |   6.060723   1.152856     5.257   0.000       3.801166    8.320281
---------+--------------------------------------------------------------------
s        |
   _cons |   .5573333   .1402154     3.975   0.000       .2825161    .8321504
------------------------------------------------------------------------------
```

When we did not specify the **gradient** option, we saw the following reported just before the **Iteration 1** line:

```
d1debug:  Begin derivative-comparison report ---------------------------------
d1debug:  mreldif(gradient vector) =  7.17e-10
d1debug:  End derivative-comparison report -----------------------------------
Iteration 1:   log likelihood = -124.10384
```

With the **gradient** option, we now see

```
Iteration 1:
d1debug:  Begin derivative-comparison report ---------------------------------
d1debug:  mreldif(gradient vector) =  7.17e-10
d1debug:  weib1-calculated gradient:
            eq1:        eq1:        eq1:        eq1:          s:
          drug2       drug3         age       _cons       _cons
r1    12.07714     11.9357    1356.576    25.82583    7.391241
d1debug:  numerically calculated gradient (used for stepping):
            eq1:        eq1:        eq1:        eq1:          s:
          drug2       drug3         age       _cons       _cons
r1    12.07714     11.9357    1356.576    25.82583    7.391241
d1debug:  relative difference:
            eq1:        eq1:        eq1:        eq1:          s:
          drug2       drug3         age       _cons       _cons
r1    2.12e-10    1.37e-11    7.17e-10    7.17e-10    1.59e-10
d1debug:  End derivative-comparison report -----------------------------------
```

We have the same **mreldif()** of 7.17e–10, but now we also see the gradient vector element by element. **mreldif()** is the maximum of the relative differences calculated element by element, where relative difference is defined as $|g_i - n_i|/(|n_i| + 1)$ and g_i is the element of the gradient vector **weib1** calculated and n_i the numerically calculated gradient value.

Specifying the **gradient** option can be useful when the calculated gradient value is suspect. In this case **weib1** works and the reported **mreldif()**s are typical:

```
iteration 0:  2.19e-03
iteration 1:  7.17e-10
iteration 2:  7.70e-10
iteration 3:  1.41e-09
iteration 4:  2.64e-05
iteration 5:  7.00e-05
iteration 6:  7.01e-05
```

What is being revealed here is not the accuracy of `weib1`, but the accuracy of the numerical approximation. The initially calculated numerical derivatives are poor, they get better, and they become poor again. They end up poor because we are measuring error as a relative difference and the derivatives themselves are going to zero.

The purpose of the output, however, is not to test the accuracy of `ml`'s numeric derivative calculator, but to verify that your analytic derivatives are in rough agreement with them. If they are, then it can be safely assumed that the analytic derivatives are correct.

To demonstrate what you might see when your analytic derivatives are wrong, we will change `weib1` to have an error. We changed the line that reads

```
mlvecsum `f' `d2' = `d' - `R'*`p'*(`M'-`d'), eq(2)
```

to read

```
mlvecsum `f' `d2' = `d' + `R'*`p'*(`M'-`d'), eq(2)
```

and so introduce a rather typical sign error. The result of estimating the model with method d1debug now is

```
. ml model d1debug weib1 (studytim died = drug2 drug3 age) /s
. ml maximize
initial:       log likelihood =       -744
alternative:   log likelihood = -356.14276
rescale:       log likelihood = -200.80201
rescale eq:    log likelihood = -136.69232
d1debug:  Begin derivative-comparison report ---------------------------------
d1debug:  mreldif(gradient vector) =  15.91092
d1debug:  End derivative-comparison report -----------------------------------
Iteration 0:   log likelihood = -136.69232  (not concave)
d1debug:  Begin derivative-comparison report ---------------------------------
d1debug:  mreldif(gradient vector) =  5.627001
d1debug:  End derivative-comparison report -----------------------------------
Iteration 1:   log likelihood = -124.10384
d1debug:  Begin derivative-comparison report ---------------------------------
d1debug:  mreldif(gradient vector) =   6.90684
d1debug:  End derivative-comparison report -----------------------------------
Iteration 2:   log likelihood = -112.12286
d1debug:  Begin derivative-comparison report ---------------------------------
d1debug:  mreldif(gradient vector) =  29.94676
d1debug:  End derivative-comparison report -----------------------------------
Iteration 3:   log likelihood = -110.29748
d1debug:  Begin derivative-comparison report ---------------------------------
d1debug:  mreldif(gradient vector) =  59.67904
d1debug:  End derivative-comparison report -----------------------------------
Iteration 4:   log likelihood = -110.26738
d1debug:  Begin derivative-comparison report ---------------------------------
d1debug:  mreldif(gradient vector) =  61.99843
d1debug:  End derivative-comparison report -----------------------------------
```

```
Iteration 5:   log likelihood = -110.26736

d1debug:  Begin derivative-comparison report ---------------------------------
d1debug:  mreldif(gradient vector) =        62
d1debug:  End derivative-comparison report -----------------------------------
Iteration 6:   log likelihood = -110.26736
                                                  Number of obs   =         48
                                                  Wald chi2(3)    =      35.25
Log likelihood = -110.26736                       Prob > chi2     =     0.0000

------------------------------------------------------------------------------
         |      Coef.   Std. Err.       z     P>|z|       [95% Conf. Interval]
---------+--------------------------------------------------------------------
eq1      |
   drug2 |   1.012966   .2903919      3.488   0.000        .4438084   1.582124
   drug3 |    1.45917   .2821196      5.172   0.000        .9062261   2.012115
     age |  -.0671728    .020569     -3.266   0.001       -.1074872  -.0268584
   _cons |   6.060723   1.152856      5.257   0.000        3.801166   8.320281
---------+--------------------------------------------------------------------
s        |
   _cons |   .5573333   .1402154      3.975   0.000        .2825161   .8321504
------------------------------------------------------------------------------
```

First note that, despite the error in **weib1**, the overall model is still estimated correctly. That is because method d1debug used the numerically calculated derivatives. The log, however, reveals that our calculation of the derivatives has problems:

```
d1debug:  Begin derivative-comparison report ---------------------------------
d1debug:  mreldif(gradient vector) =  15.91092
d1debug:  End derivative-comparison report -----------------------------------
Iteration 0:   log likelihood = -136.69232  (not concave)

d1debug:  Begin derivative-comparison report ---------------------------------
d1debug:  mreldif(gradient vector) =  5.627001
d1debug:  End derivative-comparison report -----------------------------------
Iteration 1:   log likelihood = -124.10384
```

The **mreldif()**s we now observe are

iteration 0:	15.9
iteration 1:	5.6
iteration 2:	6.9
iteration 3:	29.9
iteration 4:	59.7
iteration 5:	62.00
iteration 6:	62.00

Specifying the **gradient** option allows us to spot that the problem arises in calculating the derivative of the second equation, because we will see things like

```
Iteration 1:

d1debug:  Begin derivative-comparison report ---------------------------------
d1debug:  mreldif(gradient vector) =  5.627001

d1debug:  weib1-calculated gradient:
          eq1:       eq1:       eq1:       eq1:          s:
        drug2      drug3        age      _cons       _cons
r1   12.07714    11.9357   1356.576   25.82583   54.60876

d1debug:  numerically calculated gradient (used for stepping):
          eq1:       eq1:       eq1:       eq1:          s:
        drug2      drug3        age      _cons       _cons
r1   12.07714    11.9357   1356.576   25.82583   7.391241
```

```
d1debug:  relative difference:
            eq1:       eq1:       eq1:       eq1:          s:
          drug2      drug3        age      _cons       _cons
r1  2.12e-10  1.37e-11  7.17e-10  7.17e-10  5.627001
d1debug:  End derivative-comparison report -----------------------------------
```

Note the large values for the `s` equation and the small values for `eq1`. This tells us that the derivatives for `s` are are either derived or coded incorrectly but that the `eq1` derivatives are correct.

6.3 Robust variance estimates

If you want to specify `ml model`'s `robust` or `cluster()` options or if you want to specify `pweights`—all of which you would do on the `ml model` statement—there is something special you must do.

Here and in other parts of this book, we have said that method d0, d1, and d2 routines receive 5 arguments:

1. `` `todo' `` containing `0`, `1`, or `2`, indicating whether you are to calculate the log likelihood, the log likelihood and gradient, or log likelihood, gradient, and negative Hessian.
2. `` `b' ``, the coefficient vector.
3. `` `lnf' ``, the name of the scalar to be filled in with the log likelihood.
4. `` `g' ``, the name of a vector to be filled in with the gradient if `` `todo' `` = 1 or `` `todo' `` = 2 (and which your program may define in any case).
5. `` `negH' ``, the name of a matrix to be filled in with the negative Hessian if `` `todo' `` = 2 (and which your program may define in any case).

In fact, they receive $5+E$ arguments, where E is the number of equations (number of thetas). The E additional arguments are

6. `` `g1' ``, the name of an already existing double-precision variable to be filled in with $\partial \ln \ell_j / \partial \theta_{1j}$.
7. `` `g2' ``, the name of an already existing double-precision variable to be filled in with $\partial \ln \ell_j / \partial \theta_{2j}$.
8. ...

It is by filling in these variables that you cause `robust`, `cluster()`, and `pweights` to work.

Note that $\partial \ln \ell_j / \partial \theta_{1j}$, $\partial \ln \ell_j / \partial \theta_{2j}$, ..., are exactly what is specified on the right-hand side of the `mlvecsum`, so filling in these variables is easy enough, it is just a matter of modifying the original code,

```
program define myprog
        version 6
        args todo lnf b g
        ...
        tempname d1 d2 ...
        mlvecsum `lnf' `d1' = formula for ∂ ln ℓj/∂θ1j, eq(1)
        mlvecsum `lnf' `d2' = formula for ∂ ln ℓj/∂θ2j, eq(2)
        ...
        matrix `g' = (`d1',`d2',...)
end
```

to read

```
program define myprog
        version 6
        args todo lnf b g negH g1 g2 ...
        ...
        quietly replace `g1' = formula for ∂ ln ℓj/∂θ1j
        quietly replace `g2' = formula for ∂ ln ℓj/∂θ2j
        ...
        tempname d1 d2 ...
        mlvecsum `lnf' `d1' = `g1', eq(1)
        mlvecsum `lnf' `d2' = `g2', eq(2)
        ...
        matrix `g' = (`d1',`d2',...)
end
```

If **pweights**, **robust**, and **cluster()** are to work, **ml** needs you to fill in these variables whenever you calculate first derivatives. So here is a better version of **weib1**:

```
program define weib1
        version 6
        args todo b lnf g negH g1 g2                    /* g1 and g2 are new */
        tempvar theta1 theta2
        mleval `theta1' = `b', eq(1)
        mleval `theta2' = `b', eq(2)
        local t "$ML_y1"
        local d "$ML_y2"
        tempvar p M R
        quietly gen double `p' = exp(`theta2')
        quietly gen double `M' = (`t'*exp(-`theta1'))^`p'
        quietly gen double `R' = ln(`t')-`theta1'
        mlsum `lnf' = -`M' + `d'*(`theta2'-`theta1' + (`p'-1)*`R')
        if `todo'==0 | `lnf'==. { exit }
        quietly replace `g1' = `p'*(`M'-`d')                /* <-- new     */
        quietly replace `g2' = `d' - `R'*`p'*(`M'-`d')      /* <-- new     */
        tempname d1 d2                                      /* <-- new     */
        mlvecsum `lnf' `d1' = `g1', eq(1)                   /* <-- changed */
        mlvecsum `lnf' `d2' = `g2', eq(2)                   /* <-- changed */
        matrix `g' = (`d1',`d2')
end
```

Now we can specify the **robust** option:

```
. ml model d1 weib1 (studytim died = drug2 drug3 age) /s, robust
. ml maximize
initial:       log likelihood =       -744
alternative:   log likelihood = -356.14276
rescale:       log likelihood = -200.80201
rescale eq:    log likelihood = -136.69232
Iteration 0:   log likelihood = -136.69232  (not concave)
Iteration 1:   log likelihood = -124.13044
Iteration 2:   log likelihood =  -113.9905
Iteration 3:   log likelihood = -110.30732
Iteration 4:   log likelihood = -110.26748
Iteration 5:   log likelihood = -110.26736
Iteration 6:   log likelihood = -110.26736
```

```
                                                  Number of obs   =         48
                                                  Wald chi2(3)    =      30.69
Log likelihood = -110.26736                       Prob > chi2     =     0.0000

------------------------------------------------------------------------------
         |               Robust
         |      Coef.   Std. Err.       z     P>|z|       [95% Conf. Interval]
---------+--------------------------------------------------------------------
eq1      |
   drug2 |   1.012966   .2801112      3.616   0.000        .463958    1.561974
   drug3 |    1.45917   .2878603      5.069   0.000       .8949744    2.023366
     age |  -.0671728    .018933     -3.548   0.000      -.1042807   -.0300648
   _cons |   6.060723   1.023212      5.923   0.000       4.055264    8.066182
---------+--------------------------------------------------------------------
s        |
   _cons |   .5573333    .135934      4.100   0.000       .2909077     .823759
------------------------------------------------------------------------------
```

7 Method d2

Contents

7.1 Basic syntax

Method d2 may be used with any likelihood function,

$$\ln L = \ln L\big((\theta_{1j}, \theta_{2j}, \ldots, \theta_{Ej}; y_{1j}, y_{2j}, \ldots, y_{Dj}) : j = 1, \ldots, N\big)$$
$$\theta_{1j} = \mathbf{x}_{1j}\mathbf{b}_1$$
$$\theta_{2j} = \mathbf{x}_{2j}\mathbf{b}_2$$
$$\vdots$$
$$\theta_{Ej} = \mathbf{x}_{Ej}\mathbf{b}_E$$

where j indexes observations. The method d2 evaluator is to calculate $\ln L$, the overall log likelihood and its first and second derivatives. The outline for the program is

```
program define myprog
        version 6
        args todo b lnf g negH
        tempvar theta1 theta2 ...
        mleval `theta1´ = `b´, eq(1)
        mleval `theta2´ = `b´, eq(2) /* if there is a θ2 */
        ...
        /* if you need to create any intermediate results: */
        tempvar tmp1 tmp2 ...
        gen double `tmp1´ = ...
        ...
        mlsum `lnf´ = ...
        if `todo´==0 | `lnf´==. { exit }
        tempname d1 d2 ...
        mlvecsum `lnf´ `d1´ = formula for ∂ ln ℓj/∂θ1j , eq(1)
        mlvecsum `lnf´ `d2´ = formula for ∂ ln ℓj/∂θ2j , eq(2)
        ...
        matrix `g´ = (`d1´,`d2´,...)
        if `todo´==1 | `lnf´==. { exit }
        tempname d11 d12 d22 ...
        mlmatsum `lnf´ `d11´ = formula for −∂² ln ℓj/∂θ²1j , eq(1)
        mlmatsum `lnf´ `d12´ = formula for −∂² ln ℓj/∂θ1j∂θ2j , eq(1,2)
        mlmatsum `lnf´ `d22´ = formula for −∂² ln ℓj/∂θ²2j , eq(2)
        ...
        matrix `negH´ = (`d11´,`d12´,... \ `d12´´,`d22´,...)
end
```

1. You are passed the parameter vector. Use `mleval` to obtain the thetas from it; see **4.2.3 Using mleval**.
2. Access the dependent variables (if any) by referring to `$ML_y1`, `$ML_y2`, ... Use `$ML_y1`, `$ML_y2`, ..., just as you would any existing variable in the data.
3. The resulting log likelihood is to be saved in scalar `` `lnf' ``. Use `mlsum` to produce it; see **4.2.3 Using mleval**.
4. The resulting gradient vector is to be saved in vector `` `g' ``. Using `mlvecsum` to produce it is appropriate only when the likelihood function meets the linear-form restrictions; see **4.2.8 Using mlvecsum**.
5. The resulting *negative* Hessian matrix is to be saved in matrix `` `negH' ``. Using `mlmatsum` to produce it is appropriate only when the likelihood function meets the linear-form restrictions; see **4.2.10 Using mlmatsum**.
6. It is of great importance that, if you create any temporary variables for intermediate results, you create them as `double`s: Type "`gen double` *name* `=` ..."; do not omit the word `double`.
7. The estimation subsample is `$ML_samp==1`. You may safely ignore this if you do not specify `ml model`'s `nopreserve` option. When your program is called, only relevant data will be in memory. If you do specify `nopreserve`, understand that `mleval`, `mlsum`, `mlvecsum`, and `mlmatsum` automatically restrict themselves to the estimation subsample; it is not necessary to code `if $ML_samp==1` on these commands. You merely need to restrict other commands to the `$ML_samp==1` subsample.
8. The weights are stored in variable `$ML_w`, which contains 1 if no weights are specified. If you use `mlsum` to produce the log likelihood, `mlvecsum` to produce the gradient vector, and `mlmatsum` to produce the negative Hessian, you may ignore this because these commands handle the weights themselves.
9. While method d2 evaluators can be used with `fweights`, `aweights`, and `iweights`, you must code your program a special way if `pweights` are to be specified. You must also do this if `ml model`'s `robust` or `cluster()` options are to be specified. Change the `args` statement at the top of the program from

   ```
   args todo b lnf g negH
   ```

   to

   ```
   args todo b lnf g negH g1 g2 ...
   ```

   Change the part of the outline that reads

   ```
   tempvar d1 d2 ...
   mlvecsum `lnf' `d1' = formula for ∂ln ℓj/∂θ1j, eq(1)
   mlvecsum `lnf' `d2' = formula for ∂ln ℓj/∂θ2j, eq(2)
   ...
   ```

   to read

   ```
   quietly replace `g1' = formula for ∂ln ℓj/∂θ1j
   quietly replace `g2' = formula for ∂ln ℓj/∂θ2j
   ...
   tempvar d1 d2 ...
   mlvecsum `lnf' `d1' = `g1', eq(1)
   mlvecsum `lnf' `d2' = `g2', eq(2)
   ...
   ```

   The above will work only if the likelihood function meets the linear-form restrictions. Otherwise, `pweights`, `robust`, and `cluster()` are not available.

You specify that you are using method d2 on the `ml model` statement,

```
. ml model d2 myprog ...
```

which you issue after you have defined the likelihood-evaluation program.

7.1.1 Using method d2

The method d2 evaluator continues from where the method d1 evaluator left off. Taking out the code in common, here is how they differ:

```
program define myprog
        version 6
        args todo b lnf g negH                         /* negH is added */
        ...
        if `todo'==1 | `lnf'==. { exit }               /* from here down is new */
        tempname d11 d12 d22 ...
        mlmatsum `lnf' `d11' = formula for -∂² ln ℓj/∂θ²1j, eq(1)
        mlmatsum `lnf' `d12' = formula for -∂² ln ℓj/∂θ1j∂θ2j, eq(1,2)
        mlmatsum `lnf' `d22' = formula for -∂² ln ℓj/∂θ²2j, eq(2)
        ...
        matrix `negH' = (`d11',`d12',... \ `d12'',`d22',...)
end
```

The negative Hessian matrix you are to define is

$$-\mathbf{H} = -\frac{\partial^2 \ln L}{\partial \mathbf{b} \partial \mathbf{b}'} = \begin{pmatrix} -\frac{\partial^2 \ln L}{\partial b_1^2} & -\frac{\partial^2 \ln L}{\partial b_1 \partial b_2} & \cdots & -\frac{\partial^2 \ln L}{\partial b_1 \partial b_k} \\ -\frac{\partial^2 \ln L}{\partial b_2 \partial b_1} & -\frac{\partial^2 \ln L}{\partial b_2^2} & \cdots & -\frac{\partial^2 \ln L}{\partial b_2 \partial b_k} \\ \vdots & \vdots & \ddots & \vdots \\ -\frac{\partial^2 \ln L}{\partial b_k \partial b_1} & -\frac{\partial^2 \ln L}{\partial b_k \partial b_2} & \cdots & -\frac{\partial^2 \ln L}{\partial b_k^2} \end{pmatrix}$$

`ml` provides the coefficient vector as a $1 \times k$ row vector and expects the negative Hessian matrix to be returned as a $k \times k$ matrix.

We recommend you consider writing a method d2 evaluator only if your likelihood function meets the linear-form restrictions. In that case, adding the second derivatives is easy because `mlmatsum` will create the vector for you from the derivatives with respect to the thetas.

7.1.2 Example: Probit

We begin with probit because we want to start with a single-equation (single θ) system. The probit log-likelihood function is

$$\ln L = \sum_{j=1}^{N} \ln \ell_j$$

$$\ln \ell_j = \begin{cases} \ln \Phi(\theta_j) & \text{if } y_j = 1 \\ \ln \Phi(-\theta_j) & \text{if } y_j = 0 \end{cases}$$

$$\frac{\partial \ln \ell_j}{\partial \theta_j} = \begin{cases} \phi(\theta_j)/\Phi(\theta_j) & \text{if } y_j = 1 \\ -\phi(\theta_j)/\Phi(-\theta_j) & \text{if } y_j = 0 \end{cases}$$

$$\theta_j = \mathbf{x}_j \boldsymbol{\beta}$$

A working method d1 evaluator for probit is

```
program define myll
        version 6
        args todo b lnf g
        tempvar theta lnfj
        mleval `theta' = `b'
        quietly gen double `lnfj' = ln(normprob(`theta')) if $ML_y1==1
        quietly replace `lnfj' = ln(normprob(-`theta')) if $ML_y1==0
        mlsum `lnf' = `lnfj'
        if `todo'==0 | `lnf'==. { exit }
        tempvar gj
        quietly gen double `gj' = normd(`theta')/normprob(`theta') /*
                                */ if $ML_y1==1
        quietly replace `gj' = -normd(`theta')/normprob(-`theta') /*
                                */ if $ML_y1==0
        mlvecsum `lnf' `g' = `gj'
end
```

The negative second derivative of the probit log-likelihood function is

$$
\begin{aligned}
-\frac{\partial^2 \ln \ell_j}{\partial \theta_j^2} &= \begin{cases} R(\theta_j)\big(R(\theta_j)+\theta_j\big) & \text{if } y_j = 1 \\ S(\theta_j)\big(S(\theta_j)-\theta_j\big) & \text{if } y_j = 0 \end{cases} \\
R(\theta_j) &= \phi(\theta_j)/\Phi(\theta_j) \\
S(\theta_j) &= \phi(\theta_j)/\Phi(-\theta_j)
\end{aligned}
$$

A method d2 evaluator for this likelihood is

```
program define myll
        version 6
        args todo b lnf g negH
        tempvar theta
        mleval `theta' = `b'
        mlsum `lnf' = ln(normprob(cond($ML_y1==1,`theta',-`theta')))
        if `todo'==0 | `lnf'==. { exit }
        tempvar gj
        quietly gen double `gj' = normd(`theta')/normprob(`theta') /*
                                */ if $ML_y1==1
        quietly replace `gj' = -normd(`theta')/normprob(-`theta') /*
                                */ if $ML_y1==0
        mlvecsum `lnf' `g' = `gj'
        if `todo'==1 | `lnf'==. { exit }
        tempvar R S
        quietly gen double `R' = normd(`theta')/normprob(`theta')
        quietly gen double `S' = normd(`theta')/normprob(-`theta')
        mlmatsum `lnf' `negH'  = cond($ML_y1==1,`R'*(`R'+`theta'),`S'*(`S'-`theta'))
end
```

```
. ml model d2 myll (foreign=mpg weight)
. ml maximize
initial:       log likelihood = -51.292891
alternative:   log likelihood = -45.055272
rescale:       log likelihood = -45.055272
Iteration 0:   log likelihood = -45.055272
Iteration 1:   log likelihood = -27.904623
Iteration 2:   log likelihood = -26.858226
Iteration 3:   log likelihood = -26.844198
```

```
Iteration 4:   log likelihood = -26.844189
Iteration 5:   log likelihood = -26.844189
                                                  Number of obs   =         74
                                                  Wald chi2(2)    =      20.75
Log likelihood = -26.844189                       Prob > chi2     =     0.0000

------------------------------------------------------------------------------
 foreign |      Coef.   Std. Err.       z     P>|z|       [95% Conf. Interval]
---------+--------------------------------------------------------------------
     mpg |  -.1039503   .0515689     -2.016   0.044      -.2050235   -.0028772
  weight |  -.0023355   .0005661     -4.126   0.000       -.003445   -.0012261
   _cons |   8.275464   2.554142      3.240   0.001       3.269437    13.28149
------------------------------------------------------------------------------
```

7.1.3 Example: Weibull regression

Weibull is an example of a two-equation system. The important programming difference is that this time we will have to issue three `mlmatsum` commands (yes, three, not two) and then use `matrix` *name* = ... to put the pieces together.

The likelihood function for Weibull regression in the log-time metric is

$$
\begin{aligned}
\ln \ell_j &= (t_j e^{\theta_{1j}})^{\exp(\theta_{2j})} + d_j\big(\theta_{2j} - \theta_{1j} + (e^{\theta_{2j}} - 1)(\ln t_j - \theta_{1j})\big) \\
\theta_{1j} &= \mathbf{x}_j \mathbf{b} \\
\theta_{2j} &= s
\end{aligned}
$$

where t_j is the time of death or censoring and d_j is 1 if death and 0 if censored. The parameters to be estimated are $\mathbf{b}$ and s. It will be easier to write the likelihood evaluation programs if we rewrite the likelihood function with some intermediate variables. The log-likelihood function can be more compactly written

$$
\begin{aligned}
\ln \ell_j &= -M_j + d_j\big(\theta_{2j} - \theta_{1j} + (p_j - 1)R_j\big) \\
p_j &= e^{\theta_{2j}} \\
M_j &= (t_j e^{-\theta_{1j}})^{p_j} \\
R_j &= \ln t_j - \theta_{1j}
\end{aligned}
$$

The first derivatives of the Weibull log-likelihood function with respect to the thetas are

$$
\begin{aligned}
\frac{\partial \ln \ell_j}{\partial \theta_{1j}} &= p_j(M_j - d_j) \\
\frac{\partial \ln \ell_j}{\partial \theta_{2j}} &= d_j - R_j p_j (M_j - d_j)
\end{aligned}
$$

A working method d1 evaluator for this likelihood is

```
program define weib1
        version 6
        args todo b lnf g

        tempvar theta1 theta2
        mleval `theta1´ = `b´, eq(1)
        mleval `theta2´ = `b´, eq(2)

        local t "$ML_y1"
        local d "$ML_y2"
```

```
        tempvar p M R
        quietly gen double `p' = exp(`theta2')
        quietly gen double `M' = (`t'*exp(-`theta1'))^`p'
        quietly gen double `R' = ln(`t')-`theta1'
        mlsum `lnf' = -`M' + `d'*(`theta2'-`theta1' + (`p'-1)*`R')
        if `todo'==0 | `lnf'==. { exit }
        tempname d1 d2
        mlvecsum `lnf' `d1' = `p'*(`M'-`d'), eq(1)
        mlvecsum `lnf' `d2' = `d' - `R'*`p'*(`M'-`d'), eq(2)
        matrix `g' = (`d1',`d2')
end
```

The negative second derivatives for the Weibull model are

$$-\frac{\partial^2 \ln \ell_j}{\partial \theta_{1j}^2} = p_j^2 M_j$$

$$-\frac{\partial^2 \ln \ell_j}{\partial \theta_{1j} \partial \theta_{2j}} = -p_j(M_j - d_j + R_j p_j M_j)$$

$$-\frac{\partial^2 \ln \ell_j}{\partial \theta_{2j}^2} = p_j R_j (R_j p_j M_j + M_j - d_j)$$

To convert our working d1 evaluator into a d2 evaluator, we add these three formulas:

```
program define weib2
        version 6
        args todo b lnf g negH                          /* negH added */
        tempvar theta1 theta2
        mleval `theta1' = `b', eq(1)
        mleval `theta2' = `b', eq(2)
        local t "$ML_y1"
        local d "$ML_y2"
        tempvar p M R
        quietly gen double `p' = exp(`theta2')
        quietly gen double `M' = (`t'*exp(-`theta1'))^`p'
        quietly gen double `R' = ln(`t')-`theta1'
        mlsum `lnf' = -`M' + `d'*(`theta2'-`theta1' + (`p'-1)*`R')
        if `todo'==0 | `lnf'==. { exit }
        tempname d1 d2
        mlvecsum `lnf' `d1' = `p'*(`M'-`d'), eq(1)
        mlvecsum `lnf' `d2' = `d' - `R'*`p'*(`M'-`d'), eq(2)
        matrix `g' = (`d1',`d2')
        if `todo'==1 | `lnf'==. { exit }               /* new from here down */
        tempname d11 d12 d22
        mlmatsum `lnf' `d11' = `p'^2 * `M', eq(1)
        mlmatsum `lnf' `d12' = -`p'*(`M'-`d' + `R'*`p'*`M'), eq(1,2)
        mlmatsum `lnf' `d22' = `p'*`R'*(`R'*`p'*`M' + `M' - `d'), eq(2)
        matrix `negH' = (`d11',`d12' \ `d12'',`d22')
end
```

```
. ml model d2 weib2 (studytim died = drug2 drug3 age) /s
. ml maximize
initial:       log likelihood =       -744
alternative:   log likelihood = -356.14276
rescale:       log likelihood = -200.80201
rescale eq:    log likelihood = -136.69232
Iteration 0:   log likelihood = -136.69232  (not concave)
```

```
Iteration 1:   log likelihood = -124.13044
Iteration 2:   log likelihood = -113.99047
Iteration 3:   log likelihood = -110.30732
Iteration 4:   log likelihood = -110.26748
Iteration 5:   log likelihood = -110.26736
Iteration 6:   log likelihood = -110.26736
                                                  Number of obs   =         48
                                                  Wald chi2(3)    =      35.25
Log likelihood = -110.26736                       Prob > chi2     =     0.0000

------------------------------------------------------------------------------
         |      Coef.   Std. Err.       z     P>|z|       [95% Conf. Interval]
---------+--------------------------------------------------------------------
eq1      |
   drug2 |   1.012966   .2903917     3.488   0.000        .4438086    1.582123
   drug3 |    1.45917   .2821195     5.172   0.000        .9062261    2.012114
     age |  -.0671728   .0205688    -3.266   0.001       -.1074868   -.0268587
   _cons |   6.060723   1.152845     5.257   0.000        3.801188    8.320259
---------+--------------------------------------------------------------------
s        |
   _cons |   .5573333   .1402154     3.975   0.000        .2825163    .8321504
------------------------------------------------------------------------------
```

If we compared results with Stata's `weibull` or `streg, dist(weibull)` commands, we would note one difference: the reported value of the log likelihood is different. In the formulas above we did not include certain constants which, while having no formal role in the Weibull model, are usually included because they normalize away the units in which time are recorded. This difference is of no importance and it would just complicate our program were we to include them.

7.2 Method d2debug

We have proceeded too quickly. In real life the probit and Weibull method d2 evaluators would not have worked the first time we tried. Without a doubt we would have made an error in producing the negative Hessian, either substantively because we took the derivative incorrectly, or mechanically because we coded mistakenly.

The recipe for success is to start with a working method d1 evaluator, make the changes necessary to convert it to method d2, but then specify the method d2debug on the `ml model` statement. Method d2debug is just like method d2 except that it checks your derivatives against numerically calculated ones.

For instance, here is what happens when we test the working probit evaluator using method d2debug:

```
. ml model d2debug myll (foreign=mpg weight)
. ml maximize
initial:       log likelihood = -51.292891
alternative:   log likelihood = -45.055272
rescale:       log likelihood = -45.055272
d2debug:  Begin derivative-comparison report ---------------------------------
d2debug:  mreldif(gradient vector) =  .0002445
d2debug:  mreldif(negative Hessian) =   .000014
d2debug:  End derivative-comparison report -----------------------------------
Iteration 0:   log likelihood = -45.055272
d2debug:  Begin derivative-comparison report ---------------------------------
d2debug:  mreldif(gradient vector) =  1.02e-09
d2debug:  mreldif(negative Hessian) =  .0011529
d2debug:  End derivative-comparison report -----------------------------------
```

```
Iteration 1:   log likelihood = -27.908441
d2debug:  Begin derivative-comparison report ---------------------------------
d2debug:  mreldif(gradient vector) =  3.12e-09
d2debug:  mreldif(negative Hessian) =  .0012721
d2debug:  End derivative-comparison report -----------------------------------
Iteration 2:   log likelihood = -26.941129
d2debug:  Begin derivative-comparison report ---------------------------------
d2debug:  mreldif(gradient vector) =  .0000272
d2debug:  mreldif(negative Hessian) =  8.04e-08
d2debug:  End derivative-comparison report -----------------------------------
Iteration 3:   log likelihood = -26.845606
d2debug:  Begin derivative-comparison report ---------------------------------
d2debug:  mreldif(gradient vector) =  .0002086
d2debug:  mreldif(negative Hessian) =  1.14e-08
d2debug:  End derivative-comparison report -----------------------------------
Iteration 4:   log likelihood = -26.844189
d2debug:  Begin derivative-comparison report ---------------------------------
d2debug:  mreldif(gradient vector) =  .0004365
d2debug:  mreldif(negative Hessian) =  1.15e-08
d2debug:  End derivative-comparison report -----------------------------------
Iteration 5:   log likelihood = -26.844189
                                                  Number of obs   =         74
                                                  Wald chi2(2)    =      20.75
Log likelihood = -26.844189                       Prob > chi2     =     0.0000
------------------------------------------------------------------------------
 foreign |      Coef.   Std. Err.       z     P>|z|       [95% Conf. Interval]
---------+--------------------------------------------------------------------
     mpg |  -.1039503   .0515689    -2.016   0.044      -.2050236   -.0028771
  weight |  -.0023355   .0005661    -4.126   0.000       -.003445   -.0012261
   _cons |   8.275464   2.554144     3.240   0.001       3.269434    13.28149
------------------------------------------------------------------------------
```

The difference between methods d2 and d2debug is that method d2debug provides a derivative comparison report interspersed in the iteration log. For instance, prior to iteration 1, we see the following:

```
d2debug:  Begin derivative-comparison report ---------------------------------
d2debug:  mreldif(gradient vector) =  1.02e-09
d2debug:  mreldif(negative Hessian) =  .0011529
d2debug:  End derivative-comparison report -----------------------------------
Iteration 1:   log likelihood = -27.908441
```

Whenever `ml` needs derivatives, method d2debug calls your likelihood evaluator just as method d2 would, but method d2debug also calculates the derivatives numerically. It then displays a comparison of the derivatives calculated both ways and it returns to `ml` the *numerically calculated derivatives*. Thus, d2debug proceeds to a correct answer even if your derivative calculation (first or second) is wrong.

d2debug reports comparisons for both first and second derivatives. The issues concerning comparing first derivatives are the same as for d1debug—see **6.2 Method d1debug**—it is the value of the `mreldif()` in the middle iterations that matter.

Concerning second derivatives, it is the comparison in the last few iterations that matter. The numeric second derivatives d2debug calculates can be poor—very poor—in early and middle iterations:

```
iteration 0:  1.40e–05
iteration 1:  1.15e–03
iteration 2:  1.27e–03
iteration 3:  8.04e–08
iteration 4:  1.14e–08
iteration 5:  1.15e–08
```

That the numerical derivatives are poor in early iterations does not matter; by the last iteration the numerical derivative will be accurate, and it is with that value that you wish to compare. Remember, the purpose of the output is not to test the accuracy of `ml`'s numeric derivative calculator, but to verify that your analytic derivatives are in rough agreement with them. If they are, then it can be safely assumed that the analytic derivatives are correct.

To demonstrate what you might see when your analytic derivatives are wrong, pretend we changed our probit program to have an error. Pretend we changed the line that reads

```
quietly replace `h´ = `S´*(`S´-`theta´) if $ML_y1==0
```

to read

```
quietly replace `h´ = `S´*(`S´+`theta´) if $ML_y1==0
```

Using d2debug, our overall results would still be correct because d2debug uses the numeric derivatives (first and second), not the analytic values we calculate. But the `mreldif()`s for the Hessian would be

```
iteration 0:  .7676
iteration 1:  .6798
iteration 2:  .5499
iteration 3:  .5692
iteration 4:  .5596
iteration 5:  .5596
```

It is when the `mreldif()` value is greater than 10^{-6} in the last iteration that you should suspect problems.

When using d2debug, you can specify `ml maximize`'s `hessian` option to see the matrix compared with the numerically calculated matrix, or `gradient` to see a comparison of gradient vectors, or both.

7.3 Robust variance estimates

The issues concerning `pweights`, `robust`, and `cluster()` when using method d2 are the same as when using method d1 and so are the changes you make; see **6.3 Robust variance estimates**.

Change the `args` statement from

```
args todo b lnf g negH
```

to

```
args todo b lnf g negH g1 g2 ...
```

and then change the gradient vector calculation part of the code from

```
tempname d1 d2 ...
mlvecsum `lnf´ `d1´ = formula for ∂ln ℓ_j/∂θ_1j, eq(1)
mlvecsum `lnf´ `d2´ = formula for ∂ln ℓ_j/∂θ_2j, eq(2)
...
matrix `g´ = (`d1´,`d2´,...)
```

to

```
quietly replace `g1´ = formula for ∂ln ℓ_j/∂θ_1j
quietly replace `g2´ = formula for ∂ln ℓ_j/∂θ_2j
...
tempvar d1 d2 ...
mlvecsum `lnf´ `d1´ = `g1´, eq(1)
mlvecsum `lnf´ `d2´ = `g2´, eq(2)
...
matrix `g´ = (`d1´,`d2´,...)
```

Here is a better version of our probit routine, where we also take the opportunity to calculate the common terms in the first and second derivatives just once and so make the program execute a little faster:

```
program define myll
        version 6
        args todo b lnf g negH g1
        tempvar theta
        mleval `theta´ = `b´
        mlsum `lnf´ = ln(normprob(cond($ML_y1==1,`theta´,-`theta´)))
        if `todo´==0 | `lnf´==. { exit }
        tempvar R S
        quietly gen double `R´ = normd(`theta´)/normprob(`theta´)
        quietly gen double `S´ = normd(`theta´)/normprob(-`theta´)
        quietly replace `g1´ = cond($ML_y1==1,`R´,-`S´)
        mlvecsum `lnf´ `g´ = `g1´
        if `todo´==1 | `lnf´==. { exit }
        mlmatsum `lnf´ `negH´  = cond($ML_y1==1,`R´*(`R´+`theta´),`S´*(`S´-`theta´))
end
```

```
. ml model d2 myll (foreign=mpg weight), robust
. ml maximize
initial:       log likelihood = -51.292891
alternative:   log likelihood = -45.055272
rescale:       log likelihood = -45.055272
Iteration 0:   log likelihood = -45.055272
Iteration 1:   log likelihood = -27.904623
Iteration 2:   log likelihood = -26.858226
Iteration 3:   log likelihood = -26.844198
Iteration 4:   log likelihood = -26.844189
Iteration 5:   log likelihood = -26.844189
                                                  Number of obs   =         74
                                                  Wald chi2(2)    =      30.26
Log likelihood = -26.844189                       Prob > chi2     =     0.0000

------------------------------------------------------------------------------
         |               Robust
 foreign |      Coef.   Std. Err.       z     P>|z|       [95% Conf. Interval]
---------+--------------------------------------------------------------------
     mpg |  -.1039503   .0593548     -1.751   0.080      -.2202836    .0123829
  weight |  -.0023355   .0004934     -4.734   0.000      -.0033025   -.0013686
   _cons |   8.275464   2.539177      3.259   0.001       3.298769    13.25216
------------------------------------------------------------------------------
```

8 Debugging evaluators

Contents

8.1 Overview

It is unlikely that things will proceed as smoothly for you as they have for us in the previous chapters. You are likely to write an evaluator and observe the following:

```
. ml model lf myprobit (foreign=mpg weight)
. ml maximize
initial:       log likelihood =     -<inf>  (could not be evaluated)
--Break--
r(1);
```

We are the ones who pressed *Break* because we got tired of watching nothing happen. Or you might see

```
. ml model lf myprobit (foreign=mpg weight)
. ml maximize
invalid syntax
r(198);
```

or

```
. ml model lf myprobit (foreign=mpg weight)
. ml maximize
initial:       log likelihood =          0
alternative:   log likelihood =          0
rescale:       log likelihood =          0
could not calculate numerical derivatives
flat or discontinuous region encountered
r(430);
```

These are solvable problems and `ml` has some capabilities that will help you solve them.

8.2 ml check

The big mistake we made in the above examples was not typing `ml check` between the `ml model` and `ml maximize` statements. If you are lucky—if your program works—here is what you might see:

```
. ml model lf myprobit (foreign=mpg weight)
. ml check
Test 1:  Calling myprobit to check if it computes log likelihood and
         does not alter coefficient vector...
         Passed.
Test 2:  Calling myprobit again to check if the same log-likelihood value
         is returned...
         Passed.
Test 3:  Calling myprobit to check if 1st derivatives are computed...
         test not relevant for method lf.
Test 4:  Calling myprobit again to check if the same 1st derivatives are
         returned...
         test not relevant for method lf.
Test 5:  Calling myprobit to check if 2nd derivatives are computed...
         test not relevant for method lf.
Test 6:  Calling myprobit again to check if the same 2nd derivatives are
         returned...
         test not relevant for method lf.
--------------------------------------------------------------------------------
Searching for alternate values for the coefficient vector to verify that
myprobit returns different results when fed a different coefficient vector:
Searching...
initial:       log likelihood =     -<inf>  (could not be evaluated)
searching for feasible values +
feasible:      log likelihood = -191.66721
improving initial values +.....+...
improve:       log likelihood = -113.63338
continuing with tests...
--------------------------------------------------------------------------------
Test 7:  Calling myprobit to check log likelihood at the new values...
         Passed.
Test 8:  Calling myprobit requesting 1st derivatives at the new values...
         test not relevant for method lf.
Test 9:  Calling myprobit requesting 2nd derivatives at the new values...
         test not relevant for method lf.
--------------------------------------------------------------------------------
                         myprobit HAS PASSED ALL TESTS
--------------------------------------------------------------------------------
Test 10: Does myprobit produce unanticipated output?
         This is a minor issue.  Stata has been running myprobit with all
         output suppressed.  This time Stata will not suppress the output.
         If you see any unanticipated output, you need to place quietly in
         front of some of the commands in myprobit.
--------------------------------------------------------------- begin execution
----------------------------------------------------------------- end execution
```

It is more likely, however, that you will see something like this:

```
. ml model lf myprobit (foreign=mpg weight)
. ml check
Test 1:  Calling myprobit to check if it computes log likelihood and
         does not alter coefficient vector...
         FAILED; myprobit returned error 132.
Here is a trace of its execution:
```

```
------------------------------------------------------------------------------
-> myprobit __0005HR __0005HQ
- `begin´
- args lnf theta
- quietly replace `lnf´ = ln(normprob(`theta´)) if $ML_y1==1
- quietly replace `lnf´ = ln(normprob(-`theta´) if $ML_y1==0
too few ´)´ or ´]´
- `end´
------------------------------------------------------------------------------
Fix myprobit.
r(132);
```

We left out a close parenthesis in our program. (The problem with our program is not that it altered the coefficient vector; it was just that while attempting to test whether our program altered the coefficient vector, our program generated an error, so `ml check` backed up and showed us a trace of its execution.)

`ml check` can be used with method lf, d0, d1, or d2 evaluators.

`ml check` will save you hours of time.

8.3 Using methods d1debug and d2debug

If you have written a method d1 or d2 evaluator, start by using method d1debug or d2debug. `ml check` does not verify that your derivatives are right; it just looks for mechanical errors.

That does not mean you should not bother with `ml check`—just do not expect `ml check` to catch substantive errors.

If you have written a method d1 evaluator, we recommend that you type

```
. ml model d1debug ...
. ml check
. ml maximize
```

and if you have written a method d2 evaluator, we recommend that you type

```
. ml model d2debug ...
. ml check
. ml maximize
```

Note that we recommend use of methods d1debug and d2debug with `ml check`. They do not interact—`ml check` will produce no more information than it would with d1 or d2—but neither will `ml check` mind.

To show you what can happen, we have written a method d1 evaluator for Weibull and purposely coded the derivatives incorrectly. There are no mechanical errors in our program, however.

```
. ml model d1debug weib1 (studytim died = drug2 drug3 age) /s
. ml check
Test 1:  Calling weib1 to check if it computes log likelihood and
         does not alter coefficient vector...
         Passed.
  (output omitted)
Test 3:  Calling weib1 to check if 1st derivatives are computed...
         Passed.
(note that ml check did not notice that the derivatives are wrong!)
  (output omitted)
------------------------------------------------------------------------------
                          weib1 HAS PASSED ALL TESTS
```

```
--------------------------------------------------------------------------------
(output omitted)
You should check that the derivatives are right.
Use ml maximize to obtain estimates.
The output will include a report comparing analytic to numeric derivatives.
Do not be concerned if your analytic derivatives differ from the numeric ones
in early iterations.
The analytic gradient will differ from the numeric one in early iterations,
then the mreldif() difference should become less than 1e-6 in the middle
iterations, and the difference will increase again in the final iterations
as the gradient goes to zero.
```

`ml check` even reminds us to check that the derivatives are right.

We are already using method d1debug and we will discover our problem when we do the `ml maximize`. See **6.2 Method d1debug** for the output.

When you do run the `ml maximize` with methods d1debug or d2debug, remember to specify the `gradient` option if you want to see the gradients as well as the comparison and to specify the `hessian` option if you want to see the Hessians as well as the comparison:

```
. ml maximize, gradient
. ml maximize, hessian
. ml maximize, gradient hessian
```

In fact, we would not start that way because we do not know we have a problem yet. We would just type

```
. ml maximize
```

and then, observing that there is a problem, step back and try again:

```
. ml model d1debug weib1 (studytim died = drug2 drug3 age) /s
. ml maximize, gradient
```

8.4 ml trace

Between `ml check` and methods d1debug and d2debug, you should be able to find and fix most problems.

Even so, there may be a bug in your code that appears only rarely because it is in a part of the code that is seldomly executed. The way one typically finds such bugs is to `set trace on` and then review the output.

Do not do that with `ml` because the trace will contain a trace of `ml`'s code, too, and there is a lot of it. Instead, type `ml trace on`. That will trace the execution of your program only:

```
. ml model lf myprobit (foreign=mpg weight)
. ml trace on
. ml maximize
-> myprobit __0008SH __0008SG
- `begin'
- args lnf theta
- quietly replace `lnf' = ln(normprob(`theta')) if $ML_y1==1
- quietly replace `lnf' = ln(1-normprob(`theta')) if $ML_y1==0
- `end'
initial:       log likelihood = -51.292891
-> myprobit __0008ST __0008SS
- `begin'
- args lnf theta
```

```
- quietly replace `lnf´ = ln(normprob(`theta´)) if $ML_y1==1
- quietly replace `lnf´ = ln(1-normprob(`theta´)) if $ML_y1==0
- `end´
-> myprobit __0008T1 __0008T0
- `begin´
- args lnf theta
- quietly replace `lnf´ = ln(normprob(`theta´)) if $ML_y1==1
- quietly replace `lnf´ = ln(1-normprob(`theta´)) if $ML_y1==0
- `end´
```
(output omitted)

You may type `ml trace off` later, but that is seldom necessary because the next time you issue an `ml model` statement, trace is turned off when the previous problem is cleared.

9 Setting initial values

Contents

9.1 Overview

Typically—not always—you can get away without specifying initial values. In the examples so far in this book, we have typed

```
. ml model ...
. ml maximize
```

and bothered not at all with initial values. We have been able to do that because `ml maximize` is pretty smart about coming up with initial values on its own. By default, `ml maximize` does the following:

1. Assumes an initial value vector $\mathbf{b}_0 = (0, 0, \ldots, 0)$.
2. If $\ln L(\mathbf{b}_0)$ can be evaluated,

   a. `ml maximize` attempts to improve the initial values in a deterministic way by rescaling $\mathbf{b}_0$, overall and equation-by-equation.

   If $\ln L(\mathbf{b}_0)$ cannot be evaluated,

   b. `ml maximize` uses a pseudo-random algorithm to hunt for a $\mathbf{b}_0$ vector that will allow $\ln L(\mathbf{b}_0)$ to be evaluated and, once found,

   c. `ml maximize` attempts to improve the initial values in a deterministic way by rescaling $\mathbf{b}_0$, overall and equation-by-equation.

`ml maximize` does all of this *before* attempting to maximize the function using the standard techniques.

Actually, it is not `ml maximize` that does all of this, it is `ml search`, and `ml maximize` calls `ml search`. There are good reasons to invoke `ml search` yourself, however, and there are other ways to set the initial values, too:

1. When you invoke `ml search`, it does a more thorough job of it. When `ml search` is called by `ml maximize`, it tries to be quick and find something that is merely good enough.
2. Another way of setting initial values is `ml plot`. It graphs slices of the likelihood function (one parameter at a time) and resets the parameter to the value corresponding to the maximum of the slice. `ml plot` is great entertainment, too.

3. Another way of setting initial values is to specify them using `ml init`. This is a method of last resort when working interactively, but programmers often use this method because they are willing to work to produce good initial values. They do this because they can be faster than using either of the above techniques.

You can combine the three methods above. If you know that 5 is a good value for one of the parameters, you can use `ml init` to set it. You can then use `ml search` to grope around from there for even better values—`ml search` cannot make things worse. You can then use `ml plot` to examine another of the parameters that you know is likely to cause problems and so set it to a reasonable value. You can then use `ml search` again to improve things. If you keep working like this, you could dispense with the maximization step altogether.

9.2 ml search

The syntax of `ml search` is

> `ml search` [[/]*eqname*[:] $\#_{lb}$ $\#_{ub}$] [[/]*eqname*[:] $\#_{lb}$ $\#_{ub}$] [...]
> [, `repeat(#)` `nolog` `trace` `restart` `norescale`]

`ml search` looks for better (or feasible) initial values. By feasible, we mean values of the parameters for which the log likelihood can be calculated. For instance, in the likelihood function $\phi((y_j - \mu_j)/\sigma)$, any starting value with $\sigma = 0$ is infeasible.

The simplest syntax for `ml search` is

```
. ml search
```

Alternatively, you may follow `ml search` with the name of an equation and two numbers, such as

```
. ml search eq1 -3 3
```

and `ml search` will restrict the search to those bounds. For instance, if the likelihood function were $\phi((y_j - \mu_j)/\sigma)$ and you had typed

```
. ml model lf myprog (mu: y = x1 x2) /sigma
```

it would make sense—but not be necessary—to restrict `ml search` to positive and reasonable values for σ:

```
. ml search sigma .1 5
```

Note that you specify bounds for equations, not individual parameters. That means you do not specify bounds for the coefficient on, say, `x1`; you specify bounds for the equation `mu`. In this example, if `y` varied between 1 and 10, it would make sense to specify

```
. ml search sigma .1 5 mu 1 10
```

In practice, it is seldom worth specifying the bounds because `ml search` comes up with bounds on its own by probing your likelihood function.

Note that `ml search` defines initial values on the basis of equations. It does this by setting the intercept (coefficient on `_cons`) to numbers randomly chosen within the range and setting the remaining coefficients to zero. If the equation has no intercept, a number c within the range is chosen and the coefficients are then set from the regression $c = \mathbf{x}\mathbf{b}_0 + e$.

`ml search` may be issued repeatedly.

9.2.1 Options for ml search

`repeat(#)` specifies the number of random attempts that are to be made to find a better initial-value vector. The default is `repeat(10)`.

`repeat(0)` specifies that no random attempts are to be made. More correctly, `repeat(0)` specifies that no random attempts are to be made if the initial initial-value vector is a feasible starting point. If it is not, `ml search` will make random attempts even if you specify `repeat(0)` because it has no alternative. The `repeat()` option refers to the number of random attempts to be made to improve the initial values. When the initial starting-value vector is not feasible, `ml search` will make up to 1,000 random attempts to find starting values. It stops the instant it finds one set of values that works and then moves into its improve-initial-values logic.

`repeat(`k`)`, $k \geq 0$, specifies the number of random attempts to be made to improve the initial values.

`nolog` specifies that no output is to appear while `ml search` looks for better starting values. If you specify `nolog` and the initial starting-value vector is not feasible, `ml search` will ignore the fact that you specified the `nolog` option. If `ml search` must take drastic action to find starting values, it feels you should know about this even if you attempted to suppress its usual output.

`trace` specifies that you want more detailed output about `ml search`'s actions than it would usually provide. This is more entertaining than useful. `ml search` prints a period each time it evaluates the likelihood function without obtaining a better result and a plus when it does.

`restart` specifies that random actions are to be taken to obtain starting values and that the resulting starting values are not to be a deterministic function of the current values. Users should not specify this option mainly because, with `restart`, `ml search` intentionally does not produce as good a set of starting values as it could. `restart` is included for use by the optimizer when it gets into serious trouble. The random actions are to ensure that the actions of the optimizer and `ml search`, working together, do not result in a long, endless loop.

`restart` implies `norescale`, which is why we recommend you do not specify `restart`. In testing, cases were discovered where `rescale` worked so well that, even after randomization, the rescaler would bring the starting values right back to where they had been brought the first time and so defeated the intended randomization.

`norescale` specifies that `ml search` is not to engage in its rescaling actions to improve the parameter vector. We do not recommend specifying this option because rescaling tends to work so well.

9.2.2 Using ml search

We strongly recommend you use `ml search`. You may specify bounds for the search but that is not typically necessary. It should be enough to type

```
. ml search
```

We recommend you use `ml search` even if you specify initial values (which you do with `ml init`, and which we will explain later):

```
. ml model d2 myreg (mpg=weight displ) /sigma1
. ml init /sigma1 = 3
. ml search
```

If you use `ml search`, any previous starting values will be taken as a suggestion and it would probably be better if we typed

```
. ml model d2 myreg (mpg=weight displ) /sigma1
. ml search /sigma1 1 10
```

and so restricted the search for **/sigma** to the range $1 < \sigma < 10$ rather than suggesting $\sigma = 3$ and leaving the range at $-\infty < \sigma < \infty$. Of course, we could do both, but the suggestion is not worth nearly as much as the bounds.

Not specifying an equation does not mean the equation is not searched; it just means that no special bounds are being specified for the search.

9.2.3 Determining equation names

Remember, you do not have to specify bounds, but the search will be more efficient if you do. You specify the bounds in terms of the equations, not the individual parameters.

ml model labels equations **eq1**, **eq2**, ..., if you do not specify otherwise. You specify otherwise by coding "(*eqname*: ...)" when you specified the equation. If you are confused, you can type **ml query** anytime after the **ml model** statement to find out what you have defined and where you are. For instance, we set a problem by typing

```
. ml model lf myprobit (foreign=mpg weight)
```

and here is what **ml query** reports:

```
. ml query
Method:             lf
Program:            myprobit
Dep. variable:      foreign
1 equation:
        eq1:        mpg weight
Search bounds:
        eq1:             -inf       +inf
Current (initial) values:
        (zero)
```

Our equation is called **eq1**. Next let's search over the restricted range $(-3, 3)$:

```
. ml search eq1 -3 3
initial:       log likelihood = -51.292891
improve:       log likelihood = -47.164697
rescale:       log likelihood = -45.271144
```

If we now do an **ml query**, we see

```
. ml query
Method:             lf
Program:            myprobit
Dep. variable:      foreign
1 equation:
        eq1:        mpg weight
Search bounds:
        eq1:               -3          3
Current (initial) values:
   eq1:_cons           -.42679622
    remaining values are zero
lnL(current values) = -45.271144
```

Note that **ml** remembers the search bounds we specified. Were we now simply to type "**ml search**", the $(-3, 3)$ bounds would be assumed.

9.3 ml plot

The syntax of `ml plot` is

`ml plot` [*eqname*:]*name* [# [# [#]]]

`ml plot` graphically assists in filling in initial values. In practice, this command is cute more than useful because `ml search` works so well. Still, the focus of `ml plot` is on the individual parameters, not the overall equation as with `ml search`, and that can sometimes be useful.

`ml plot` graphs the likelihood function for one parameter at a time and then replaces the value of the parameter according to the maximum value of the likelihood function. These graphs are of cuts of the likelihood function—they are not graphs of the profile likelihood function.

Understand that `ml plot` is not an alternative to `ml search`; you can use them both, in any order, and repeatedly. We recommend you use `ml search` first and only then try `ml plot`. This way, `ml plot` is at least starting with feasible values.

The simplest syntax is to type `ml plot` followed by a coefficient name:

```
. ml plot _cons
. ml plot /sigma
. ml plot eq2:_cons
. ml plot eq1:foreign
```

When the name of the equation is not explicitly supplied, the first equation, whatever the name, is assumed (thus, `_cons` probably means `eq1:_cons`).

For instance, after defining,

```
. ml model lf myprobit (foreign=mpg weight)
```

we might type

```
. ml plot _cons
                reset _cons =         -.5  (was          0)
             log likelihood = -45.055272  (was -51.292891)
```

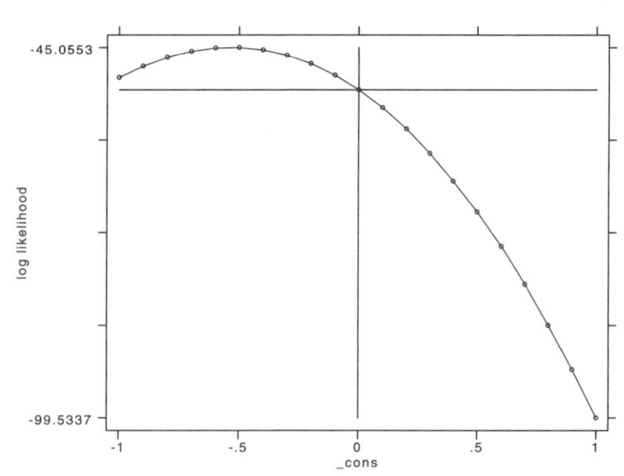

By default, `ml plot` graphs 21 values of the likelihood within ± 1 of the current value of the parameter. Rather than displaying a graph, `ml plot` might present you with a message:

```
. ml plot _cons
(all values are missing)
```

This could mean

1. The current value for the coefficient on `_cons` and all values between ±1 of the current value are infeasible; you need to find a better range.
2. The current value for the coefficient on `_cons` is infeasible and the range is too wide. Remember, `ml plot` graphs 21 points; perhaps values within ±1/100th of the range would be feasible.

That is why we recommend you first use `ml search`.

Alternatively, you can type `ml plot` followed by the name of a parameter followed by a number, such as

```
. ml plot weight .01
                keeping weight =          0
                log likelihood = -45.055272
```

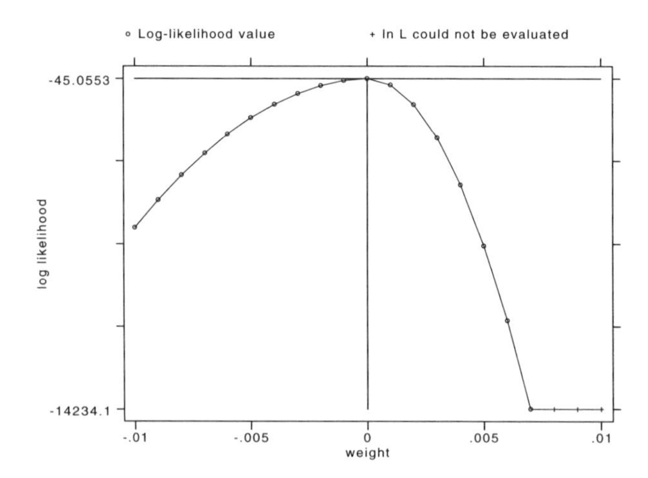

In this case we graph ±.01 around the current value of the coefficient on `weight`. If we specified two numbers, such as "`ml plot mpg -.5 1`", we would graph the range $[b-.5, b+1]$ where b would be the current value of the specified parameter. If we typed a third number, that many points (plus 1) would be graphed.

The ml system maintains the current value of the initial-value vector $\mathbf{b}_0$. $\mathbf{b}_0$ is set to zero when we issue an `ml model` statement and thereafter, other routines can update it. Each time we perform an `ml search`, $\mathbf{b}_0$ is updated. Each time we use `ml plot`, $\mathbf{b}_0$ is updated if a better value for the parameter is found. Thus, you can use `ml search` and `ml plot` together. The current value of $\mathbf{b}_0$ is reported by `ml query`.

9.4 ml init

The syntax of `ml init` is

`ml init` { [*eqname*:]*name*=# | /*eqname*=# } [...]

`ml init` # [# ...], `copy`

`ml init` *matname* [, `skip copy`]

`ml init` allows you explicitly to specify initial values. `ml init` without arguments resets the parameter vector to contain all zeros.

9.4.1 Options for mlinit

`skip` specifies that any parameters found in the specified initialization vector that are not also found in the model are to be ignored. The default action is to issue an error message.

`copy` specifies that the initialization vector or list of numbers is to be copied into the initial-value vector by position rather than by name.

9.4.2 Using ml init

`ml init` may be issued not at all, once, or repeatedly.

To specify initial values, you type the coefficient's name, an equal sign, and the value to which it should be set. For instance,

```
. ml model lf myprob (foreign=mpg weight)
. ml init mpg=1.5 _cons=-.5
```

Note that more than one coefficient may be specified and not all coefficients need be specified. Initial values for omitted coefficients remain unchanged.

In multiple-equation models, specify the equation name and a colon to identify the coefficient:

```
. ml model lf myprob (y1=x1 x2 x3) (y2=x1 x4) /sigma1 /sigma2
. ml init eq1:_cons=5 eq2:_cons=3
```

In this case, the first two equations are named **eq1** and **eq2** because we did not explicitly specify equation names in the `ml model` statement. When you omit the equation name in the `ml init` statement, as in

```
. ml init x1=3.14
```

it is assumed you are referring to the first equation.

In addition, when setting the coefficient on the constant, you may refer to *eqname*`:_cons` or to `/`*eqname*—they mean the same thing:

```
. ml init /sigma1=3 sigma2:_cons=4
```

You may even use the slash notation with equations that have more than just a constant. The line

```
. ml init eq1:_cons=5 eq2:_cons=3
```

could just as well be specified

```
. ml init /eq1=5 /eq2=3
```

In the rare instance where you wish to specify initial values for all parameters, you can type them one after the other:

```
. ml init 5 7 2 9  3 2 1  1  1, copy
```

In most such cases, however, you would have a vector containing the values and it is easier to use the syntax described below.

9.4.3 Loading initial values from vectors

Initial values may be obtained from row vectors (matrices). This is rarely done and this feature is included for programmers who want to obtain initial values from some other estimator:

```
. regress y1 x2 x3
. matrix b0 = e(b)
. ml model lf myprob (y1=x1 x2 x3) (y2=x1 x4) /sigma1 /sigma2
. ml init b0
```

By default, values are copied from the specified vector by name: it does not matter that the vector obtained from `regress` is in a different order than that used by `ml`. In this case the initial values were applied to the first equation because the vector `b0` fetched from `regress` did not include equation names; had we set the equation names, the vector could have applied to the second equation.

Our use of the name `b0` has no special significance in the above example; we could just as well have called it `initvec` or anything else.

10 Interactive maximization

Contents

10.1 Overview

You type `ml maximize` to obtain estimates once you have issued the `ml model` statement. The process can be as short as

```
. ml model ...
. ml maximize
```

but we recommend

```
. ml model ...
. ml check
. ml search
. ml maximize
```

10.2 The iteration log

If you type `ml maximize`, you will see something like

```
. ml maximize
initial:       log likelihood =       -744
alternative:   log likelihood = -356.14276
rescale:       log likelihood = -200.80201
rescale eq:    log likelihood = -136.69232
Iteration 0:   log likelihood = -136.69232  (not concave)
Iteration 1:   log likelihood = -124.13044
Iteration 2:   log likelihood =  -113.9905
...
```

The first three lines of the output really have to do with `ml search`. The first line, `initial`, reports the log likelihood at the initial values $\mathbf{b}_0$. In this case the log likelihood is -744, but it might be reported as `-<inf>`, meaning the likelihood could not be evaluated at $\mathbf{b}_0$. That would not cause problems. The next line, `alternative`, reports the result of a random search for an alternative $\mathbf{b}_0$.

ml then takes the better of the two. `rescale` reports the results of rescaling the entire $\mathbf{b}_0$ vector; that is, search for a scalar c such that $\ln L(c\mathbf{b}_0) > \ln L(\mathbf{b}_0)$. `rescale eq` then repeats the search for c, equation by equation.

Understand that maximization—the modified Newton–Raphson technique—has not even begun. All that is happening is that starting values are being improved, and already we have improved the log likelihood from -744 to -136.7.

The goal is to find a $\mathbf{b}_0$ such that the Newton–Raphson method has a reasonable chance of converging to the maximum.

It is at iteration 0 that the maximization process really starts, and note that iteration 0 reports an opening log likelihood that is equal to where the searcher left off.

The purpose of an iteration is to take $\mathbf{b}_i$ and produce a better value $\mathbf{b}_{i+1}$, where i is the iteration number. Iteration 0 takes $\mathbf{b}_0$ and produces $\mathbf{b}_1$.

An iteration is defined as a calculation of a new direction vector. When `ml` reports an iteration, it is reporting $\ln L(\mathbf{b}_i)$, the value received. The iteration calculates a direction vector $\mathbf{d}_i$ based on the function at that point: $\mathbf{d}_i$ = some function of the gradient and the Hessian. The iteration then steps in that direction until the log likelihood ceases to improve. $\mathbf{b}_{i+1}$ is defined as $\mathbf{b}_i + s\mathbf{d}_i$, where s is the step size taken. The process then repeats.

Thus, first and second derivatives are calculated once per iteration. Between times, log-likelihood evaluations are made.

10.3 Pressing the Break key

You may press *Break* while `ml` is iterating.

```
. ml maximize

initial:       log likelihood =       -744
alternative:   log likelihood = -356.14276
rescale:       log likelihood = -200.80201
rescale eq:    log likelihood = -136.69232
Iteration 0:   log likelihood = -136.69232  (not concave)
Iteration 1:   log likelihood = -124.13044
Iteration 2:   log likelihood =  -113.9905
Iteration 3:   log likelihood = -110.30732
--Break--
r(1);

. _
```

Nothing is lost; you can pick up right from where you left off:

```
. ml maximize

initial:       log likelihood = -110.30732
rescale:       log likelihood = -110.30732
rescale eq:    log likelihood = -110.30732
Iteration 0:   log likelihood = -110.30732
Iteration 1:   log likelihood = -110.26748
Iteration 2:   log likelihood = -110.26736
(estimation output appears)
```

The iteration numbers change, but note the opening log-likelihood values are the same as the values just before pressing *Break*.

In between, you can use `ml query` to obtain a summary of where you are,

```
. ml query

Method:           d1
Program:          weib1
Dep. variables:   studytim died
2 equations:
        eq1:      drug2 drug3 age
        /s
Search bounds:
        eq1:           -inf       +inf
        /s             -inf       +inf
```

```
Current (initial) values:
   eq1:drug2            1.01373
   eq1:drug3          1.4600912
   eq1:age           -.06716715
   eq1:_cons          6.0590723
   s:_cons            .55862111
lnL(current values) = -110.30732
```

or type `ml report` for a fuller report:

```
. ml report
Current coefficient vector:
          eq1:        eq1:        eq1:        eq1:          s:
         drug2       drug3         age       _cons       _cons
r1    1.01373    1.460091   -.0671672    6.059072    .5586211
Value of log-likelihood function = -110.26748
Gradient vector (length = 5.771333):
          eq1:        eq1:        eq1:        eq1:          s:
         drug2       drug3         age       _cons       _cons
r1   .0029079    .0004267    5.769681     .101467   -.0936065
Negative Hessian matrix (concave; matrix is full rank):
                    eq1:        eq1:        eq1:        eq1:          s:
                   drug2       drug3         age       _cons       _cons
eq1:drug2       18.34357
eq1:drug3              0    18.33923
  eq1:age        1082.52    1047.396    308919.7
eq1:_cons       18.34357    18.33923    5392.432    94.92623
  s:_cons       5.649744    5.499115   -321.9273   -4.897996     60.8065
Steepest-ascent direction:
          eq1:        eq1:        eq1:        eq1:          s:
         drug2       drug3         age       _cons       _cons
r1   .0005038    .0000739    .9997137    .0175812   -.0162192
Newton-Raphson direction (length before normalization = .0024007):
          eq1:        eq1:        eq1:        eq1:          s:
         drug2       drug3         age       _cons       _cons
r1  -.3170628   -.3821464   -.0023107    .6840327   -.5343493
```

You may even use any of the methods for setting initial values—`ml search`, `ml plot`, or `ml init`—to attempt to improve the parameter vector before continuing with the maximization by typing `ml maximize`.

10.4 Maximizing difficult likelihood functions

When you have a difficult likelihood function, specify `ml maximize`'s `difficult` option. `difficult` varies how a direction is found when the negative Hessian $-\mathbf{H}$ cannot be inverted and thus the usual direction calculation $\mathbf{d} = \mathbf{g}(-\mathbf{H})^{-1}$ cannot be calculated. `ml` flags such instances by mentioning that the likelihood is "not concave":

```
. ml maximize
initial:       log likelihood =       -744
alternative:   log likelihood = -356.14276
rescale:       log likelihood = -200.80201
rescale eq:    log likelihood = -136.69232
Iteration 0:   log likelihood = -136.69232  (not concave)
Iteration 1:   log likelihood = -124.13044
Iteration 2:   log likelihood =  -113.9905
...
```

`ml`'s usual solution is to mix a little steepest ascent into the calculation and that works well enough in most cases.

When you specify `difficult`, `ml` goes to more work. `ml` computes the eigenvalues of $-\mathbf{H}$ and then, for the part of the orthogonal subspace where the eigenvalues are negative or small positive numbers, uses steepest ascent and in the other subspace uses a regular Newton–Raphson step.

Understand that the presence of a "not concave" message does not necessarily flag problems. Typically it is merely maddening when `ml` reports "not concave" iteration after iteration as it crawls toward a solution. It is only a problem if `ml` reports "not concave" at the last iteration. The following log has "not concave" messages along the way and at the end:

```
. ml maximize
initial:       log likelihood =  -2702.155
alternative:   log likelihood =  -2702.155
rescale:       log likelihood = -2315.2187
rescale eq:    log likelihood = -2220.5288
Iteration 0:   log likelihood = -2220.5288  (not concave)
Iteration 1:   log likelihood = -2152.5211  (not concave)
Iteration 2:   log likelihood = -2141.3916  (not concave)
Iteration 3:   log likelihood = -2140.7298  (not concave)
Iteration 4:   log likelihood = -2139.5807  (not concave)
Iteration 5:   log likelihood = -2139.0407  (not concave)
  (output omitted)
Iteration 40:  log likelihood = -2138.0806  (not concave)
Iteration 41:  log likelihood = -2138.0803  (not concave)
Iteration 42:  log likelihood = -2138.0801  (not concave)
Iteration 43:  log likelihood = -2138.0799  (not concave)
(results presented; output omitted)
```

Specifying `difficult` sometimes helps:

```
. ml maximize, difficult
initial:       log likelihood =  -2702.155
alternative:   log likelihood =  -2702.155
rescale:       log likelihood = -2315.2187
rescale eq:    log likelihood = -2220.5288
Iteration 0:   log likelihood = -2220.5288  (not concave)
Iteration 1:   log likelihood = -2143.5267  (not concave)
Iteration 2:   log likelihood = -2138.2358
Iteration 3:   log likelihood = -2137.9355
Iteration 4:   log likelihood = -2137.8971
Iteration 5:   log likelihood = -2137.8969
(results presented; output omitted)
```

In this case, `difficult` not only sped estimation but solved the problem of nonconcavity at the final iteration. Note that the final log likelihood without `difficult` was −2138.1 which is less than −2137.9. The difference may seem small, but it was enough to place the final results on a concave portion of the likelihood function.

Usually, however, specifying `difficult` just slows things down.

11 Final results

Contents

11.1 Overview

Once you have typed **ml maximize** and obtained the results, you may do any of the things Stata usually allows after estimation. That means you can use **test** to perform hypotheses tests, **predict** to obtain predicted values (the linear predicted values from each of the equations), and the like.

11.2 Graphing convergence

A special feature of **ml** is that, once estimation output has been produced, you may graph the convergence by typing **ml graph**.

```
. ml model d1 weibl (studytim died = drug2 drug3 age) /s
. ml maximize
initial:       log likelihood =       -744
alternative:   log likelihood = -356.14276
rescale:       log likelihood = -200.80201
rescale eq:    log likelihood = -136.69232
Iteration 0:   log likelihood = -136.69232  (not concave)
Iteration 1:   log likelihood = -124.13044
Iteration 2:   log likelihood =  -113.9905
Iteration 3:   log likelihood = -110.30732
Iteration 4:   log likelihood = -110.26748
Iteration 5:   log likelihood = -110.26736
Iteration 6:   log likelihood = -110.26736
                                                  Number of obs   =         48
                                                  Wald chi2(3)    =      35.25
Log likelihood = -110.26736                       Prob > chi2     =     0.0000

------------------------------------------------------------------------------
         |      Coef.   Std. Err.       z     P>|z|       [95% Conf. Interval]
---------+--------------------------------------------------------------------
eq1      |
   drug2 |   1.012966   .2903917      3.488   0.000       .4438087    1.582123
   drug3 |    1.45917   .2821195      5.172   0.000       .9062261    2.012114
     age |  -.0671728   .0205687     -3.266   0.001      -.1074868   -.0268587
   _cons |   6.060723   1.152845      5.257   0.000       3.801189    8.320258
---------+--------------------------------------------------------------------
s        |
   _cons |   .5573333   .1402153      3.975   0.000       .2825163    .8321504
------------------------------------------------------------------------------
```

```
. ml graph
```

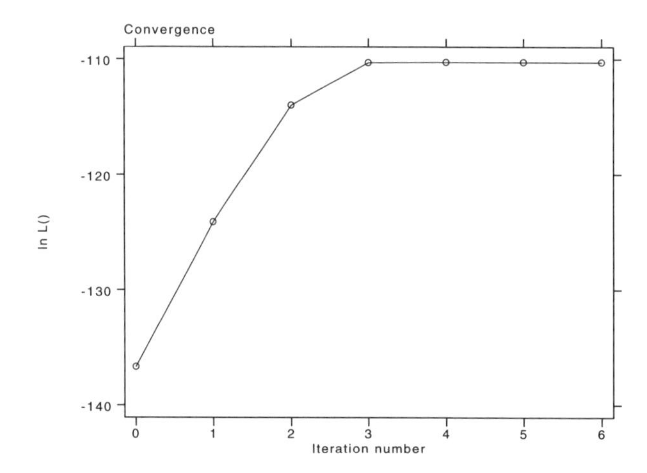

Given how Newton–Raphson optimizers work, for smooth likelihood functions the graph may jump around at first but, once it gets close to the maximum, it should move smoothly and with smaller and smaller steps toward it; see **1.5 Monitoring convergence**. This graph shows that the process has truly converged.

The syntax of `ml graph` is

ml graph [#]

specifies the number of iterations to be graphed, counted from the last. "`ml graph 5`" would graph the last five iterations.

`ml graph` keeps track of the last 20 iterations; this is sufficient for determining convergence.

11.3 Redisplaying output

Once `ml maximize` has displayed results, you may use `ml display` to redisplay it.

```
. ml display

                                                  Number of obs   =         48
                                                  Wald chi2(3)    =      35.25
Log likelihood = -110.26736                       Prob > chi2     =     0.0000

------------------------------------------------------------------------------
         |      Coef.   Std. Err.       z     P>|z|       [95% Conf. Interval]
---------+--------------------------------------------------------------------
eq1      |
   drug2 |   1.012966   .2903917      3.488   0.000       .4438087    1.582123
   drug3 |    1.45917   .2821195      5.172   0.000       .9062261    2.012114
     age |  -.0671728   .0205687     -3.266   0.001      -.1074868   -.0268587
   _cons |   6.060723   1.152845      5.257   0.000       3.801189    8.320258
---------+--------------------------------------------------------------------
s        |
   _cons |   .5573333   .1402153      3.975   0.000       .2825163    .8321504
------------------------------------------------------------------------------
```

The syntax of `ml display` is

`ml display [, eform(string) noheader first neq(#) plus level(#) ]`

The options are

`eform(`*string*`)` specifies that the exponentiated form of the coefficients is to be reported; *string* is used to label the exponentiated coefficients.

`noheader` suppresses display of the standard header above the coefficient table.

`first` restricts the display of the coefficient table to the first equation and also suppresses displaying the name of the first equation.

`neq(`*#*`)` restricts the display of the coefficient table to the first # equations; equation names are shown.

`plus` displays the coefficient table just as it would be ordinarily, but then, rather than ending the table in a line of dashes, ends it in dashes–plus-sign–dashes. This is so that programmers can write additional display code to add more results to the table and make it appear as if the combined result is one table. Programmers typically specify `plus` with options `first` or `neq()`.

`level(`*#*`)` specifies the confidence level for confidence intervals of the coefficients.

12 Writing do-files to maximize likelihoods

Contents

12.1 When to use do-files

There are two reasons you maximize likelihood functions:

1. You need to estimate the model for a particular analysis, or just a few analyses.
2. You need the estimator around constantly; you use it all the time.

We recommend using do-files in case (1) and ado-files (covered in the next chapter) in case (2).

12.2 Structure of do-file

We organize our do-files differently depending on the problem, but we always start the same way:

——— top of outline.do ———

```
clear

capture program drop myprog
program define myprog
        version 6
        args ...
        ...
end

use sampledata
ml model ... myprog ...
```

——— end of outline.do ———

Then, working interactively, we can type

```
. do outline
. ml check
. ml search
. ml max
```

Typically, `ml check` detects problems, but this structure makes it easy to try again. We can edit our do-file and then start all over by typing "`do outline`". We do not actually name the file `outline.do`, of course. At this stage we typically give it some very short name: `t.do` is our favorite. Then we can type "`do t`".

Notice that we put the loading of some sample data and the `ml model` statement in the do-file. The remaining `ml` commands we need to type are short; we just did not want to be forever retyping things like

```
. ml model d2 weib2 (studytim died = drug2 drug3 age) /s
```

while we are debugging our code.

12.3 Putting the do-file into production

Once we have our evaluator working, we modify our do-file and rename it:

——— top of *readytouse.do* ———

```
capture program drop myprog
program define myprog
        version 6
        args ...
        ...
end

capture program drop model
program define model
        version 6
        ml model  method myprog (`0´) /s
end
```

——— end of *readytouse.do* ———

The changes we made were

1. We removed the `clear` at the top; we no longer want to eliminate the data in memory. Now we can load the data and the do-file in any order.
2. We removed the "`use` *sampledata*" and the hardcoded `ml model` statement.
3. We substituted in their place a new program—`model`. `model` is a two-line program we write to make it easy to type the `ml model` statement.

We previously imagined that model statements such as

```
. ml model d2 weib2 (studytim died = drug2 drug3 age) /s
```

were reasonable for this estimator. Note that some of what we need to type will always be the same:

```
. ml model d2 weib2 (...) /s
```

To make it easier for us to use the estimator, we code the fixed part into a new program:

```
program define model
        version 6
        ml model d2 weib2 (`0´) /s
end
```

`` `0´ `` is just Stata jargon for "what the user typed goes here", so with `model` we can type things like

```
. model studytim died = drug1 drug2 age
```

With *readytouse*`.do` in place, in our analysis do-file, we can code things like

——— top of analysis.do ———

```
run readytouse
use realdata
model studytim died = drug1 drug2 age
ml search
ml max
```

——— end of analysis.do ———

13 Writing ado-files to maximize likelihoods

Contents

13.1 Writing estimation commands

The term estimation command has a special meaning in Stata as it implies a number of properties:

1. Estimation commands share a common syntax.
2. You can, at any time, review the previous estimates by typing the estimation command without arguments.
3. Estimation commands share a common output format.
4. You can obtain the variance–covariance matrix of the estimators by typing "`vce`" after estimation.
5. You can restrict subsequent commands to the estimation subsample by suffixing them with `if e(sample)`, for example "`summarize ... if e(sample)`".
6. You can obtain predictions and the standard error of the linear prediction using `predict`.
7. You can refer to coefficients and standard errors in expressions (such as with `generate`) by typing `_b[`*name*`]` or `[`*eqname*`]_b[`*name*`]`.
8. You can perform tests on the estimated parameters using `test` (Wald test of linear hypotheses), `nltest` (Wald test of nonlinear hypotheses), and `lrtest` (likelihood-ratio tests). You can obtain point estimates and confidence intervals for linear combinations of estimated parameters using `lincom`.

Except for items 1 and 2, `ml` satisfies all of these requirements; you do not have to do anything more. `ml`'s syntax is different from other estimation commands (item 1) and you cannot type `ml` without arguments to redisplay results (item 2), but you can type `ml display` to redisplay results, so the second violation is on a technicality.

The primary reason to package an `ml` estimator into an ado-file is convenience of syntax. Rather than typing

```
. do such-and-such (to load the evaluator)
. ml model lf weibll (studytim died = drug2 drug3 age) /s if age>20, robust
. ml search
. ml maximize
```

it would be more convenient simply to type

```
. myweib studytim died drug2 drug3 age if age>20, robust
```

Creating such an ado-file would not be difficult:

——— top of myweib.ado ———

```
program define myweib
        version 6
        syntax varlist(min=2) [if] [in] [, Robust]
        tokenize `varlist'
        local lhs1 "`1'"
        local lhs2 "`2'"
        mac shift 2
        local rhs "`*'"
        ml model lf weibll (`lhs1' `lhs2'=`rhs') /s `if' `in', `robust'
        ml search
        ml maximize
end
```

——— end of myweib.ado ———

——— top of weibll.ado ———

```
program define weibll
        version 6
        args lnf theta1 theta2
        tempvar p M R
        quietly gen double `p' = exp(`theta2')
        quietly gen double `M' = ($ML_y1*exp(-`theta1'))^`p'
        quietly gen double `R' = ln($ML_y1)-`theta1'
        quietly replace `lnf' = -`M' + $ML_y2*(`theta2'-`theta1' + (`p'-1)*`R')
end
```

——— end of weibll.ado ———

The above program works although it does not really meet our standards of an estimation command. Typing `myweib` without arguments will not redisplay results, although it would not be difficult to incorporate that feature. `myweib` allows a `robust` option, but not a `cluster()` option. That too could be easily fixed, but perhaps you never use the `cluster()` option.

The point is `myweib` works and it may work well enough for your purposes. We are about to go to considerable effort to solve what are really minor criticisms. We want to add every conceivable option and we want to make the thing bulletproof. Such work is justified only if you or others plan on using the resulting command often.

13.2 The standard estimation-command outline

The outline for any estimation command—not just those written with `ml`—is

top of *newcmd*.ado

```
program define newcmd, eclass
        version 6
        local options "display-options"
        if replay() {
                if "`e(cmd)'" ~= "newcmd" {
                        error 301       /* last-estimates-not-found error */
                }
                syntax [, `options']
        }
        else {
                syntax ... [, estimation-options `options' ]
                /* identify estimation subsample */
                marksample touse
                /* obtain the estimation results */
                ...
                estimate local cmd "newcmd"
        }
        display estimation results respecting display options
end
```

end of *newcmd*.ado

The idea behind this outline is

1. Estimation commands should redisplay results when typed without arguments.
2. Estimation commands should produce and display results when typed with arguments.

Variations on the outline are acceptable and a favorite one is

top of *newcmd*.ado

```
program define newcmd, eclass
        version 6
        if replay() {
                if "`e(cmd)'" ~= "newcmd" {
                        error 301
                }
                Replay `0'
        }
        else    Estimate `0'
end

program define Estimate, eclass
        syntax ... [, estimation-options display-options ]
        /* identify estimation subsample */
        marksample touse
        /* obtain the estimation results */
        ...
        estimate local cmd "newcmd"
        Replay, ...
end

program define Replay
        syntax [, display-options ]
        display estimation results respecting the display options
end
```

end of *newcmd*.ado

The names `Estimate` and `Replay` are actually used for the subroutine names; there is no possibility of conflict even if other ado-files use the same names because the subroutines are local to this ado-file. (Note that we do not bother to restate "`version 6`" on the subroutines `Estimate` and `Replay`. The subroutines can only be called by *newcmd*, and *newcmd* is already declared to be `version 6`.)

We will use this second outline. Obviously all that is involved is to substitute `ml` as a subroutine used in producing the estimates. Nevertheless, there are some issues about using `ml` as a subroutine, and so we include some `ml` options we have not shown previously and we add a few details to the outline that we would add to any estimation command, such as allowing for a `level()` option.

13.3 Outline for estimation commands using ml

The outline for a new estimation command implemented with `ml` is

top of *newcmd*.ado

```
program define newcmd, eclass
        version 6
        if replay() {
                if "`e(cmd)'" ~= "newcmd" {
                        error 301
                }
                Replay `0'
        }
        else    Estimate `0'
end

program define Estimate, eclass
        syntax varlist [if] [in] [fweight pweight] [, /*
                */ Robust CLuster(varname) SCore(passthru) /*
                */ other-estimation-options /*
                */ Level(passthru) other-display-options-if-any ]
        if "`cluster'"~="" {
                local clopt "cluster(`cluster')"
        }
        /* identify estimation subsample */
        marksample touse
        markout `touse' `cluster', strok
        /* perform estimation using ml */
        display in green _n "Fitting constant-only model:"
        ml model method myll ... [`weight'`exp'] if `touse', /*
                */ maximize search(quietly) nopreserve
        display in green _n "Fitting full model:"
        ml model method myll ... [`weight'`exp'] if `touse', /*
                */ continue maximize search(off) nopreserve /*
                */ `robust' `clopt' `score'
        estimate local cmd "newcmd"
        Replay, `level' other-display-options-if-any
end

program define Replay
        syntax [, Level(int $S_level) other-display-options-if-any ]
        ml display, level(`level')
end
```

end of *newcmd*.ado

top of *myll*.ado

```
program define myll
        version 6
        args ...
        ...
end
```

end of *myll*.ado

Note that an estimation command implemented in terms of `ml` becomes two ado-files:

newcmd`.ado`	The estimation command the user sees
myll`.ado`	The secret subroutine (the evaluator) `ml` calls

The evaluator must be in a separate ado-file because otherwise `ml` could not find it.

We recommend the secret subroutine be given the name formed by the first five characters of *newcmd* suffixed with `_ll`: *newcm*`_ll`. No user will ever have to type this name, nor would we want the user to type it, and placing an underscore in the name pretty well ensures that. The standard ending `_ll` reminds you that it is a log-likelihood evaluator. Using the first five characters from the name of the command reminds you with which ado-file this subroutine goes.

For instance, if you planned on calling your new estimation command `myprobit`, then we recommend you name the likelihood evaluator `mypro_ll.ado`.

13.4 Using ml in noninteractive mode

The second thing to note about our outline is that there are no `ml search`, `ml init`, or `ml maximize` commands. We use only `ml model` and then, later, `ml display`.

`ml model`, it turns out, has a `maximize` option and `ml model` can issue the `ml maximize` command for us. It also has a `search()` option corresponding to `ml search`, an `init()` option corresponding to `ml init`, and a lot of other options besides.

The entire `ml` system has two modes of operation, interactive mode—which we have used in the previous chapters—and noninteractive mode—which we want to use now. When you specify `ml model`'s `maximize` option, you turn off interactive mode. The `ml model` command proceeds from definition to final solution.

Turning off interactive mode is desirable in this context. We wish to use `ml` as a subroutine and we do not want the user to be aware of that. If we used `ml` in interactive mode—if we coded the `ml model` and `ml maximize` statements separately—then if the user pressed *Break* at the wrong time, `ml` would leave behind traces of the incomplete maximization. In addition, in interactive mode, `ml` temporarily adds variables to the dataset, but these variables are not official temporary variables as issued by the `tempvar` command. Instead, `ml` gets rid of them when the `ml maximize` command successfully completes. These variables could be left behind were the user to press *Break*.

In noninteractive mode, `ml` uses real temporary variables and arranges that all traces of itself disappear no matter how things work out.

Noninteractive mode is turned on—interactive mode is switched off—when we specify `ml model`'s `maximize` option.

13.5 ml model

Here is the full syntax of `ml model`:

> `ml model` *method progname eq* [*eq* ...] [*weight*] [`if` *exp*] [`in` *range*] [`,`
> *standard_options*
> *advanced_options*
> *interactive_options*
> *noninteractive_option*
> *noninteractive_initial_value_options*
> *noninteractive_search_options*
> *noninteractive_maximize_options*]

where *standard_options* are
: `robust cluster(`*varname*`) title(`*string*`)`

and *advanced_options* are
: `nopreserve collinear missing lf0(`$\#_k$ $\#_{ll}$`) continue waldtest(`*#*`)`
`obs(`*#*`) noscvars`

and *interactive_options* are
: `noclear`

and *noninteractive_option* is
: `maximize`

and *noninteractive_initial_value_options* are
: `continue init(`*ml_init_args*`)`

and *noninteractive_search_options* are
: `search(on|quietly|off) repeat(`*#*`) bounds(`*ml_search_bounds*`)`

and *noninteractive_maximization_options* are
: `difficult nolog trace gradient hessian showstep`
`iterate(`*#*`) ltolerance(`*#*`) tolerance(`*#*`) nowarning`
`novce score(`*newvarlist*`)`

13.5.1 Standard options

`robust` and `cluster(`*varname*`)` specify the robust estimate of variance, as does specifying `pweights`.

If you have written a method lf evaluator, `robust`, `cluster()`, and `pweights` will work. There is nothing to do except specify the options.

If you have written a method d0 evaluator, `robust`, `cluster()`, and `pweights` will not work. Specifying these options will result in an error message.

If you have written a method d1 or d2 evaluator, `robust`, `cluster()`, and `pweights` will work if you fill in the scores—see **6.3 Robust variance estimates**—otherwise specifying these options will result in an error message.

`title(`*string*`)` specifies the title to be placed on the estimation output when results are complete.

13.5.2 Advanced options

`nopreserve` specifies that it is not necessary for `ml` to ensure that only the estimation subsample is in memory when the user-written likelihood evaluator is called. `nopreserve` is irrelevant when using method lf.

For the other methods, if `nopreserve` is not specified, then `ml` saves the dataset in a file (preserves the original dataset) and drops the irrelevant observations before calling the user-written evaluator. This way, even if the evaluator is written sloppily and does not restrict its attentions to the `$ML_samp==1` subsample, results will still be correct. (Later, `ml` automatically restores the original data.)

`ml` need not go through these machinations in the case of method lf because the user-written evaluator calculates observation-by-observation values and it is `ml` itself that sums the components.

`ml` goes through these machinations if and only if the estimation sample is a subsample of the data in memory. If the estimation sample includes every observation in memory, `ml` does not preserve the original data. Thus, programmers must not damage the original data; they must `preserve` the data themselves if that is their intention.

We recommend interactive users of `ml` not specify `nopreserve`; the speed gain is not worth the chances of incorrect results.

We recommend programmers do specify `nopreserve`, but only after verifying that their evaluator really does restrict its attentions solely to the `$ML_samp==1` subsample.

`collinear` specifies that `ml` is not to remove collinear variables within equations. There is no reason that one would want to leave collinear variables in place, but this option is of interest to programmers who, in their code, have already removed collinear variables and thus do not want `ml` to waste computer time checking again.

`missing` specifies that observations containing variables with missing values are not to be eliminated from the estimation sample. There are two reasons one might want to specify `missing`.

Programmers may wish to specify `missing` because, in other parts of their code, they have already eliminated observations with missing values and thus do not want `ml` to waste computer time looking again.

All users may wish to specify `missing` if their model explicitly deals with missing values. Stata's `heckman` command is a good example of this. In such cases, there will be observations where missing values are allowed and other observations where they are not—where their presence should cause the observation to be eliminated. If you specify `missing`, it is your responsibility to specify an `if` *exp* that eliminates the irrelevant observations.

`lf0(`$\#_k$ $\#_{ll}$`)` is typically used solely by programmers. It specifies the number of parameters and log-likelihood value of the "constant-only" model so that `ml` can report a likelihood-ratio test rather than a Wald test. These values were, perhaps, analytically determined, or they may have been determined by a previous estimation of the constant-only model on the estimation sample.

Also see the `continue` option directly below.

If you specify `lf0()`, it must be safe for you to specify the `missing` option, too, else how did you calculate the log likelihood for the constant-only model on the same sample? You must have identified the estimation sample, and done so correctly, so there is no reason for `ml` to waste time rechecking your results. If you are wrong, the reported likelihood-ratio test will be wrong. Which is to say, do not specify `lf0()` unless you are certain your code identifies the estimation sample correctly.

lf0(), even if specified, is ignored if robust, cluster(), or pweights are specified because in that case a likelihood-ratio test would be inappropriate.

continue is typically specified solely by programmers. It does two things:

First, it specifies that a model has just been estimated, by either ml or some other estimation command such as logit, and that the likelihood value stored in e(ll) and the number of parameters in e(b) that are stored at this instant are the relevant values of the constant-only model. The current value of the log likelihood is ignored if robust, cluster(), or pweights are specified because in that case a likelihood-ratio test would be inappropriate.

Second, continue sets the initial values $\mathbf{b}_0$ for the model about to be estimated according to the e(b) currently stored. This is overridden if you specify any other noninteractive-initial-value options.

The comments made about specifying missing with lf0() apply equally well in this case.

waldtest(#) is typically specified solely by programmers. By default, ml presents a Wald test, but that is overridden if options lf0() or continue are specified, and that is overridden again (so we are back to the Wald test) if robust, cluster(), or pweights are specified.

waldtest(0) prevents even the Wald test from being reported.

waldtest(-1) is the default. It specifies that a Wald test is to be performed, if it is performed, by constraining all coefficients except for the intercept to 0 in the first equation. Remaining equations are to be unconstrained. The logic as to whether a Wald test is performed is the standard: perform the Wald test if neither lf0() nor continue were specified, but force a Wald test if robust, cluster, or pweights were specified.

waldtest(k) for $k \leq -1$ specifies that a Wald test is to be performed, if it is performed, by constraining all coefficients except for intercepts to 0 in the first $|k|$ equations; remaining equations are to be unconstrained. The logic as to whether a Wald test is performed is the standard.

waldtest(k) for $k \geq 1$ works like the above except that it forces a Wald test to be reported even if the information to perform the likelihood-ratio test is available and even if none of robust, cluster, or pweights were specified. waldtest(k), $k \geq 1$, may not be specified with lf0().

obs(#) is used mostly by programmers. It specifies that the number of observations reported, and ultimately stored in e(N), is to be #. Ordinarily, ml works that out for itself, and correctly. Programmers may want to specify this option when, in order for the likelihood-evaluator to work for N observations, they first had to modify the data so that it contained a different number of observations.

noscvars is used mostly by programmers. It specifies that method d0, d1, or d2 is being used but that the likelihood evaluation program does not calculate nor use arguments `g1´, `g2´, etc., which are the score vectors; see **6.3 Robust variance estimates**. Thus, ml can save a little time by not bothering to generate or pass those arguments.

13.5.3 Interactive options

noclear specifies that later, when the ml maximize command is given, and has converged, the ml problem definition is not to be cleared. Perhaps the user is having convergence problems and intends to run the model to convergence, then use ml search to see if those values can be improved, and then start the estimation again.

noclear may not be specified in noninteractive mode.

13.5.4 Noninteractive option

`maximize` specifies noninteractive mode; `ml model` is to proceed from definition, through maximization, to the posting of final results. If `maximize` is specified, the only other `ml` commands available are `ml display` (for displaying output) and `ml graph` (for graphing convergence).

If `maximize` is not specified, `ml` works in interactive mode. The model is defined and the initial value vector $\mathbf{b}_0$ is set to 0, but no other actions are taken.

13.5.5 Noninteractive initial-value options

`continue` was described above in **13.5.2 Advanced options**. Use of `continue` does not require that `maximize` also be specified.

`continue` does two things, one having to do with recording `e(ll)` from the current model so that a likelihood-ratio test can later be performed, and the second being to set the initial values $\mathbf{b}_0$ based on the currently estimated `e(b)`.

If you want the initial-value feature but not the likelihood-ratio feature, you specify `continue` with either `lf0()` or `waldtest()`; see **13.5.2 Advanced options** above.

If you want the likelihood-ratio feature but not the initial value feature, specify `continue` and the `init()` option, described directly below. `init()` overrides `continue` as far as initial values are concerned.

`init(`*ml_init_args*`)` sets the initial values $\mathbf{b}_0$. *ml_init_args* are whatever you would type after the `ml init` command and thus, the following would all be valid:

```
init(mpg=1.5 _cons=0.5)
init(eq1:_cons=5 eq2:_cons=3)
init(/eq1=5 /eq2=3)
init(b0)                  (where b0 is the name of a vector)
init(b0, skip)
```

See **9.4 ml init**.

13.5.6 Noninteractive search options

`search(on|quietly|off)` specifies whether `ml search` is to be used to improve the initial values.

`search(on)` is the default. By default, the search performed is equivalent to "`ml search, repeat(0)`" were the model being estimated in interactive mode. Thus, the search is deterministic if the likelihood function can be evaluated at the initial values $\mathbf{b}_0$, which is $(0, 0, \ldots, 0)$ or as previously set by `init()`. If the likelihood cannot be evaluated, random actions are taken until a $\mathbf{b}_0$ is found and then that is subjected to rescaling, both overall and equation by equation.

`search(quietly)` is the same in terms of results as `search(on)`; it is equivalent to `ml search repeat(0) nolog`" in interactive mode: The log of the search which would appear above the iteration log is suppressed unless random actions are required.

`ml search(off)` suppresses the running of `ml search` altogether. The initial values $\mathbf{b}_0$ are used as is. If the likelihood cannot be evaluated at $\mathbf{b}_0$, the error message "initial values not feasible", r(1400), will be issued and no model estimated.

`repeat(`*#*`)` is relevant only if `search(off)` is not specified. It specifies the number of random attempts to be made to improve the initial values $\mathbf{b}_0$.

`bounds(`*ml_search_bounds*`)` is relevant only if `search(off)` is not specified and `repeat(#)`, # > 0, is specified. It sets the bound for the random search. The syntax for specifying bounds is the same as it is with the `ml search` command: *eqname* $\#_{lb}$ $\#_{ub}$, with bounds listed one after the other.

13.5.7 Noninteractive maximization options

`difficult` specifies that the likelihood function is likely to be difficult to maximize. In particular, `difficult` states that there may be regions where $-\mathbf{H}$ is not invertible and that, in those regions, `ml`'s standard fixup may not work well. `difficult` specifies that a different fixup requiring substantially more computer time is to be used. For the majority of likelihood functions, `difficult` is likely to increase execution times unnecessarily. For other likelihood functions, specifying `difficult` is of great importance.

`nolog`, `trace`, `gradient`, `hessian`, and `showstep` control the display of the iteration log.

`nolog` suppresses reporting of the iteration log.

`trace` adds to the iteration log a report on the current parameter vector.

`gradient` adds to the iteration log a report on the current gradient vector.

`hessian` adds to the iteration log a report on the current negative Hessian matrix.

`showstep` adds to the iteration log a report on the steps within iteration.

`iterate(#)`, `ltolerance(#)`, and `tolerance(#)` specify the definition of convergence.

`iterate(16000) tolerance(1e-6) ltolerance(1e-7)` are the default.

Convergence is declared when

$$\begin{aligned} & \texttt{mreldif}(\mathbf{b}_{i+1}, \mathbf{b}_i) \le \texttt{tolerance()} \\ \textbf{or} \quad & \texttt{reldif}(\ln L(\mathbf{b}_{i+1}), \ln L(\mathbf{b}_i)) \le \texttt{ltolerance()} \end{aligned}$$

In addition, iteration stops when i = `iterate()`; in that case, results along with the message "convergence not achieved" are presented. The return code is still 0, however.

`nowarning` is allowed only with `iterate(0)`. `nowarning` suppresses the "convergence not achieved" message. Programmers might specify `iterate(0) nowarning` when they have a vector **b** already containing what are the final estimates and want `ml` to calculate the variance matrix and post final estimation results. In that case, specify "`init(b) search(off) iterate(0) nowarning nolog`".

`novce` is allowed only with `iterate(0)`. `novce` substitutes the zero matrix for the variance matrix which in effect posts estimation results as fixed constants.

`score(`*newvarlist*`)` specifies that the equation scores are to be stored in the specified new variables. Either specify one new variable name per equation or specify a short name suffixed with a `*`. E.g., `score(sc*)` would be taken as specifying `sc1` if there were one equation and `sc1` and `sc2` if there were two equations. In order to specify `score()`, either you must be using method lf or the estimation subsample must be the entire data in memory or you must have specified the `nopreserve` option (see **13.5.2 Advanced options**).

13.6 ml display

Use `ml display` to display final estimation results after use of `ml model, maximize` (use of `ml model` in noninteractive mode). The syntax of `ml display` is

`ml display [, noheader eform(string) first neq(#) plus level(#) ]`

The options are

`noheader` suppresses display of the header above the coefficient table that displays the final log-likelihood value, the number of observations, and the model-significance test.

`eform(`*string*`)` displays the coefficient table in exponentiated form: for each coefficient, $\exp(b)$ rather than b is displayed and standard errors and confidence intervals are transformed. Display of the intercept, if any, is suppressed. *string* is the table header that will be displayed above the transformed coefficients and must be 11 characters or less in length, for example, `eform("Odds ratio")`

`first` displays a coefficient table reporting results for the first equation only, and the report makes it appear that the first equation is the only equation. This is used by programmers who estimate ancillary parameters in the second and subsequent equations and report the values of such parameters themselves.

`neq(#)` is an alternative to `first`. `neq(#)` displays a coefficient table reporting results for the first # equations. This is used by programmers who estimate ancillary parameters in the $\# + 1$ and subsequent equations and report the values of such parameters themselves.

`plus` displays the coefficient table just as it would be ordinarily, but then, rather than ending the table in a line of dashes, ends it in dashes–plus-sign–dashes. This is so that programmers can write additional display code to add more results to the table and make it appear as if the combined result is one table.

`level(#)` is the standard confidence-level option. It specifies the confidence level, in percent, for confidence intervals of the coefficients.

13.7 Advice

13.7.1 Syntax

Keep the syntax simple. Remember, you have to parse it and the user has to type it. Making the syntax unnecessarily complicated serves nobody's interest.

Many models are just single equations with one or more "dependent" variables. In that case, we recommend the syntax be

cmd varlist [*weight*] [`if` *exp*] [`in` *range*] [, ...]

Note that no extraneous characters are placed between the dependent and independent variables. This syntax is easy to parse and use. If there is one dependent variable,

```
syntax varlist [fweight pweight] [if] [in] [, ... ]
tokenize `varlist'
local lhs "`1'"
mac shift
local rhs "`*'"
...
ml model ... (`lhs'=`rhs') ...
```

Note that this works even if there is really more than one equation but the subsequent equations concern ancillary parameters such as the variance of the residual in linear regression or the shape parameter in Weibull regression. The `ml model` statement would read

```
ml model ... (`lhs´=`rhs´) /sigma ...
```

If there are two dependent variables (e.g., Stata's **intreg** command), the outline becomes

```
syntax varlist(min=2) [fweight pweight] [if] [in] [, ... ]
tokenize `varlist´
local lhs1 "`1´"
local lhs2 "`2´"
mac shift 2
local rhs "`*´"
...
ml model ... (`lhs1´ `lhs2´=`rhs´) ...
```

Many multiple-equation models lend themselves to the second equation being specified as a required option. Stata's **heckman** command is an example of this. In those cases, the outline (with one dependent variable) becomes

```
syntax varlist [fweight pweight] [if] [in], SECond(varlist) [... ]
tokenize `varlist´
local lhs "`1´"
mac shift
local rhs "`*´"
...
ml model ... (`lhs´=`rhs´) (`second´)  ...
```

Above we specified the independent variables for the second equation in the required option `second()`, minimal abbreviation `sec()`. The name of the option, of course, would be of your choosing.

13.7.2 Estimation subsample

In your code you may need to make calculations on the data. If so, you must ensure that you make it on the same data that `ml` uses, which is to say, the estimation subsample.

We strongly recommend you identify that sample early in your code and then restrict your attentions to it. In the standard Stata syntax case, this is easy:

```
syntax varlist(min=#) [fweight pweight] [if] [in] [, ... ]
marksample touse
```

Thereafter you code **if `touse´** on the end of your Stata commands. (**marksample** creates a temporary variable—a variable you subsequently refer to in single quotes—and that variable contains 1 if the observation is to be used and 0 otherwise.)

If you obtain a second equation from an option, do not forget to account for missing values in it, as well:

```
syntax varlist [fweight pweight] [if] [in], SECond(varlist) [, ... ]
marksample touse
markout `touse´ `second´
```

markout takes the **`touse´** variable created by **marksample** and sets elements in it to zero in observations where the second, third, ..., specified variables are missing.

In our outline for an estimation command using `ml`, we allowed for the `cluster(`*varname*`)` option. `cluster()` substitutes the robust variance calculation and further informs it that observations should not be considered independent; merely the clusters identified by *varname* are independent. *varname* itself could contain missing values, so we would need to add

```
syntax ... [, CLuster(varname) ... ]
marksample touse
markout `touse' `cluster', strok
```

`ml` allows the clustering variable to be a string identifier. The `strok` option on `markout` tells `markout` not to think of strings as missing values. Thus, if we put this together with a second equation specified by a required option, we would have

```
syntax ... , SECond(varlist) [CLuster(varname) ... ]
marksample touse
markout `touse' `second'
markout `touse' `cluster', strok
```

Once we have correctly identified the estimation subsample, there is no reason for `ml` to go to the work of doing it again. We would specify our `ml model` statement as

```
ml model ... if `touse', missing ...
```

`ml model`'s `missing` option literally means that missing values among the dependent and independent variable are okay and so not to mark them out from the estimation sample. In fact, we have already marked such missing values out, so we are merely saving time. We are also ensuring that, if we have marked the estimation subsample wrongly, `ml` will use the same incorrect sample and that will probably cause the estimation to go violently wrong. Thus a bug will not go undiscovered.

13.7.3 Weights

We recommend that estimation commands allow frequency weights and sampling weights if they can. Concerning the latter, the issues are the same as for allowing the robust estimates of variance. With method lf, there is no issue; with method d0, it is not possible; and with methods d1 and d2, you must write your likelihood evaluator in a special way; see **6.3 Robust variance estimates**.

If you allow weights, we recommend that you code

```
syntax ... [fweight pweight] ...
...
if "`weight'" ~= "" {
        tempvar wvar
        qui gen double `wvar' `exp' if `touse'
        local weight "[`weight'=`wvar']
}
```

The idea here is to evaluate the weighting expression once and store it in a temporary variable and to redefine macro `` `weight' `` so that it is more convenient to use. Subsequently, you can code the `ml model` statement as

```
. ml model ... `weight' if `touse' ...
```

because either `` `weight' `` substitutes to nothing or it substitutes to `[fweight=`*tmpvar*`]` or `[pweight=`*tmpvar*`]`. Moreover, you now have `` `wvar' `` which you can use should that be needed.

13.7.4 Constant-only model

We recommend that you adopt a two-step procedure for obtaining estimates:

```
display in green _n "Fitting constant-only model:"
ml model ..., maximize ...
display in green _n "Fitting full model:"
ml model ..., maximize continue search(off) ...
```

The idea here is first to fit the "constant-only" model and then to proceed to the full model. There are two reasons we recommend this:

1. Users want to see reported a likelihood-ratio test if it is relevant. The two-step procedure, with the `continue` option used at the second step, will do this.
2. Estimation of many models proceeds more smoothly if one starts with $\mathbf{b}_0$ filled in from the constant-only model.

With some likelihood functions you can analytically obtain a solution for the constant-only model. If so, do that. The outline then becomes

```
obtain analytic solution
display in green _n "Fitting full model:"
ml model ..., maximize search(off) lf0(...) init(...) ...
```

Note our use of `search(off)` in the second step in both outlines. We do that because it is exceedingly unlikely that `ml search` will improve on these starting values and so searching would be a waste of time. Consider omitting `search(off)` only if the constant-only model involves setting any coefficients on `_cons` to zero in any equation or if some equations have no `_cons`.

13.7.5 Initial values

In the previous chapters we have been very lax about initial values and we have been able to be so because `ml` is good about filling them in for us. In estimation commands, however, we recommend against using this approach.

Mostly, `ml search` does a good job and, if randomly it should not, the user will type `ml search` yet again. In programming a canned estimator, do not depend on randomness working in your favor.

We recommend that you produce known-to-be-good starting values. For most models, it is adequate to produce known-to-be-good starting values for the constant-only model, obtain the constant-only estimates, and then proceed from there to the full estimates:

```
produce good starting values for constant-only model
display in green _n "Fitting constant-only model:"
ml model ..., maximize init(...) search(off) ...
display in green _n "Fitting full model:"
ml model ..., maximize continue search(off) ...
```

It would be acceptable to remove the `search(off)` from the first `ml model` statement. If, however, your starting values really are good, you want to leave the `search(off)` in place because otherwise you will just be wasting computer time.

If you had questionable starting values, in addition to removing `search(off)` you might be tempted to add, say, `repeat(5)`, and thus allow `ml search` a better opportunity to improve the values. In general we do not do this because, if randomly searching really does improve things, then we are going to have to explain to users why they got one iteration log one time and a different iteration log another. We would also be worried that randomness may not work in our favor next time. This would all be an argument to work harder on the problem of initial values so that randomly searching really is a waste of computer time.

Notice our very different attitude when programming an estimator for our own use and writing an estimation command. When we are just using `ml` interactively, we seldom bother with initial values. We use `ml search` and that is safe because we are monitoring results. When we write an estimation command, users may not anticipate problems and therefore not watch for them. Because the estimator is a command, they will assume it works, and we therefore recommend performing the work necessary to meet their expectations.

13.7.6 Saving results in e()

`ml` automatically saves the standard things in `e()` for you. You need to add `e(cmd)` so that the rest of Stata will know these results were produced by a full-fledged estimation command:

```
. est local cmd "newcmd"
```

You may want to store other things in `e()` for your own use, which is to say, for use by your `Replay` subroutine. Store them before setting `e(cmd)` so that, should the user press *Break* while you are still setting values in `e()`, Stata will know that the estimation is not complete.

Do not store anything in `e()` until after you have issued the last `ml model` command. `ml model` wipes out whatever is stored in `e()` every time it is run.

13.8 Example: Probit

Here is an example of a single-equation estimator using probit coded doing everything right. The constant-only model in probit is $b_0 = \Phi^{-1}(\bar{p})$ where $\bar{p}$ is the average number of positive outcomes observed in the estimation sample. Even though it is not necessary, we are going to use b_0 as the initial value in a first round of estimation just so you can see how that is done.

——— top of myprobit.ado ———

```
program define myprobit, eclass
        version 6
        if replay() {
                if "`e(cmd)'" ~= "myprobit" {
                        error 301
                }
                Replay `0'
        }
        else    Estimate `0'
end

program define Estimate, eclass
        syntax varlist [fweight pweight] [if] [in] [, /*
                */ Robust CLuster(varname) Level(passthru) ]
        if "`cluster'" ~= "" {
                local clopt "cluster(`cluster')"
        }
        tokenize `varlist'
        local lhs "`1'"
        mac shift
        local rhs "`*'"
                                        /* define estimation sample     */
        marksample touse
        markout `touse' `cluster', strok
                                        /* handle weights               */
        if "`weight'" ~= "" {
                tempvar wvar
                gen double `wvar' `exp' if `touse'
                local weight "[`weight'=`wvar']"
        }
                                        /* produce initial value        */
        if "`weight'" ~= "" {
                qui summarize `lhs' [aw=`wvar'] if `touse', meanonly
        }
        else    qui summarize `lhs' if `touse', meanonly
        local b0 = invnorm(r(mean))
                                        /* estimate constant-only model */
```

```
        display in green _n "Fitting constant-only model:"
        ml model lf mypro_ll (`lhs'=) `weight' if `touse', max miss /*
               */ init(_cons=`b0') search(off) nopreserve
                                        /* estimate full model         */
        display in green _n "Fitting full model:"
        ml model lf mypro_ll (`lhs'=`rhs') `weight' if `touse', max miss /*
               */ continue search(off) nopreserve /*
               */ `robust' `clopt' title("My probit")
                                        /* fill in e()                 */
        est local cmd "myprobit"
                                        /* display final results       */
        Replay, `level'
end

program define Replay
        syntax [, Level(int $S_level)]
        ml display, level(`level')
end
```
——— end of myprobit.ado ———

——— top of mypro_ll.ado ———
```
program define mypro_ll
        version 6
        args lnf theta
        quietly replace `lnf' = ln(normprob(`theta')) if $ML_y1==1
        quietly replace `lnf' = ln(normprob(-`theta')) if $ML_y1==0
end
```
——— end of mypro_ll.ado ———

Here is `myprobit` in use:

```
. myprobit foreign mpg weight
Fitting constant-only model:
Iteration 0:   log likelihood =  -45.03321
Iteration 1:   log likelihood =  -45.03321
Fitting full model:
Iteration 0:   log likelihood =  -45.03321
Iteration 1:   log likelihood = -27.914626
Iteration 2:   log likelihood = -26.857228
Iteration 3:   log likelihood = -26.844196
Iteration 4:   log likelihood = -26.844189
Iteration 5:   log likelihood = -26.844189
My probit                                         Number of obs   =         74
                                                  LR chi2(2)      =      36.38
Log likelihood = -26.844189                       Prob > chi2     =     0.0000
------------------------------------------------------------------------------
 foreign |      Coef.   Std. Err.       z     P>|z|       [95% Conf. Interval]
---------+--------------------------------------------------------------------
     mpg |  -.1039503   .0515689     -2.016   0.044      -.2050235   -.0028772
  weight |  -.0023355   .0005661     -4.126   0.000       -.003445   -.0012261
   _cons |   8.275464   2.554142      3.240   0.001       3.269438    13.28149
------------------------------------------------------------------------------
```

We wrote `myprobit` in terms of method lf. Had we instead used method d2, we would change the two `ml model` lines in `myprobit.ado` from reading

```
ml model lf ...
```

to

```
ml model d2 ...
```

and we would substitute for mypro_ll.ado:

top of mypro_ll.ado

```
program define mypro_ll
        version 6
        args todo b lnf g negH g1
        tempvar theta
        mleval `theta' = `b'
        mlsum `lnf' = ln(normprob(cond($ML_y1==1,`theta',-`theta')))
        if `todo'==0 | `lnf'==. { exit }
        tempvar R S
        quietly gen double `R' = normd(`theta')/normprob(`theta')
        quietly gen double `S' = normd(`theta')/normprob(-`theta')
        quietly replace `g1' = cond($ML_y1==1,`R',-`S')
        mlvecsum `lnf' `g' = `g1'
        if `todo'==1 | `lnf'==. { exit }
        mlmatsum `lnf' `negH'  = cond($ML_y1==1,`R'*(`R'+`theta'),`S'*(`S'-`theta'))
end
```

end of mypro_ll.ado

A Syntax diagrams

Syntax

```
ml clear

ml model      method progname eq [eq ...] [weight] [if exp] [in range] [,
              robust cluster(varname) title(string) nopreserve
              collinear missing lf0(#_k #_ll) continue waldtest(#)
              obs(#) noscvars ]

ml query

ml check

ml search     [ [/]eqname[:] #_lb #_ub ] [...]
              [, repeat(#) nolog trace restart norescale ]

ml plot       [eqname:]name [# [# [#]]] [, saving(filename[, replace]) ]

ml init       { [eqname:]name=# | /eqname=# } [...]
ml init       # [# ...], copy
ml init       matname [, skip copy]

ml report

ml trace      { on | off }

ml count      [ clear | on | off ]

ml maximize   [, difficult nolog trace gradient hessian showstep
              iterate(#) ltolerance(#) tolerance(#) nowarning novce
              score(newvarnames) nooutput level(#) eform(string) noclear ]

ml graph      [#] [, saving(filename[, replace]) ]

ml display    [, noheader eform(string) first neq(#) plus level(#) ]
```

In the preceding, *method* is { `lf` | `d0` | `d1` | `d1debug` | `d2` | `d2debug` }.

eq is the equation to be estimated, enclosed in parentheses, and optionally with a name to be given to the equation, preceded by a colon:

`(`[*eqname*`:`] [*varnames* `=`] [*varnames*] [`,` *eq_options*]`)`

or *eq* is the name of a parameter such as sigma with a slash in front

`/`*eqname* which is equivalent to `(`*eqname*`:)`

and *eq_options* are

`noconstant` `offset(`*varname*`)` `exposure(`*varname*`)`

`fweights`, `pweights`, `aweights`, and `iweights` are allowed, see [U] **14.1.6 weight** in the *Stata User's Guide*. With all but method lf, you must write your likelihood-evaluation program a certain way if `pweights` are to be specified, and `pweights` may not be specified with method d0.

`ml` shares the features of all estimation commands; see [U] **23 Estimation and post-estimation commands** in the *Stata User's Guide*. To redisplay results, type `ml display`.

Syntax of ml model in noninteractive mode

`ml model` *method progname eq* [*eq* ...] [*weight*] [`if` *exp*] [`in` *range*] `, maximize`

[`robust` `cluster(`*varname*`)` `title(`*string*`)` `nopreserve` `collinear`

`missing` `lf0(`$\#_k$ $\#_{ll}$`)` `continue` `waldtest(`*#*`)` `obs(`*#*`)` `noscvars`

`init(`*ml_init_args*`)` `search(on` | `quietly` | `off)` `repeat(`*#*`)`

`bounds(`*ml_search_bounds*`)` `difficult` `nolog` `trace` `gradient`

`hessian` `showstep` `iterate(`*#*`)` `ltolerance(`*#*`)` `tolerance(`*#*`)`

`nowarning` `novce` `score(`*newvarlist*`)`]

Noninteractive mode is invoked by specifying option `maximize`. Use `maximize` when `ml` is to be used as a subroutine of another ado-file or program and you want to carry forth the problem, from definition to posting of final results, in one command.

Syntax of subroutines for use by method d0, d1, and d2 evaluators

`mleval` *newvarname* `=` *vecname* [`,` `eq(`*#*`)`]

`mleval` *scalarname* `=` *vecname*`,` `scalar` [`eq(`*#*`)`]

`mlsum` $scalarname_{\text{lnf}}$ `=` *exp* [`if` *exp*] [`,` `noweight`]

`mlvecsum` $scalarname_{\text{lnf}}$ *rowvecname* `=` *exp* [`if` *exp*] [`,` `eq(`*#*`)`]

`mlmatsum` $scalarname_{\text{lnf}}$ *matrixname* `=` *exp* [`if` *exp*] [`,` `eq(`*#*[`,`*#*]`)`]

Syntax of user-written evaluator

Summary of notation

The log-likelihood function is $\ln L(\theta_{1j}, \theta_{2j}, \ldots, \theta_{Ej})$ where $\theta_{ij} = \mathbf{x}_{ij}\mathbf{b}_i$ and $j = 1, \ldots, N$ indexes observations and $i = 1, \ldots, E$ indexes the linear equations defined by `ml model`. If the likelihood satisfies the linear-form restrictions, it can be decomposed as $\ln L = \sum_{j=1}^{N} \ln\ell(\theta_{1j}, \theta_{2j}, \ldots, \theta_{Ej})$.

Method lf evaluators:

```
program define progname
        version 6
        args lnf theta1 [theta2 ... ]
        /* if you need to create any intermediate results: */
        tempvar tmp1 tmp2 ...
        quietly gen double `tmp1' = ...
        ...
        quietly replace `lnf' = ...
end
```

where

`lnf'	variable to be filled in with observation-by-observation values of $\ln \ell_j$
`theta1'	variable containing evaluation of 1st equation $\theta_{1j}=\mathbf{x}_{1j}\mathbf{b}_1$
`theta2'	variable containing evaluation of 2nd equation $\theta_{2j}=\mathbf{x}_{2j}\mathbf{b}_2$

See Appendix B, **Method lf checklist**.

Method d0 evaluators:

```
program define progname
        version 6
        args todo b lnf
        tempvar theta1 theta2 ...
        mleval `theta1' = `b', eq(1)
        mleval `theta2' = `b', eq(2) /* if there is a θ2 */
        ...
        /* if you need to create any intermediate results: */
        tempvar tmp1 tmp2 ...
        gen double `tmp1' = ...
        ...
        mlsum `lnf' = ...
end
```

where

`todo'	always contains 1 (may be ignored)
`b'	full parameter row vector $\mathbf{b}=(\mathbf{b}_1,\mathbf{b}_2,\ldots,\mathbf{b}_E)$
`lnf'	scalar to be filled in with overall $\ln L$

See Appendix C, **Method d0 checklist**.

Method d1 evaluators:

```
program define progname
        version 6
        args todo b lnf g [negH g1 [g2 ... ] ]
        tempvar theta1 theta2 ...
        mleval `theta1' = `b', eq(1)
        mleval `theta2' = `b', eq(2) /* if there is a θ2 */
        ...
```

```
        /* if you need to create any intermediate results: */
        tempvar tmp1 tmp2 ...
        gen double `tmp1´ = ...
        ...
        mlsum `lnf´ = ...
        if `todo´==0 | `lnf´==. { exit }
        tempname d1 d2 ...
        mlvecsum `lnf´ `d1´ = formula for ∂ ln ℓj/∂θ1j, eq(1)
        mlvecsum `lnf´ `d2´ = formula for ∂ ln ℓj/∂θ2j, eq(2)
        ...
        matrix `g´ = (`d1´, `d2´, ... )
end
```

where

`` `todo´ ``	contains 0 or 1 0⇒`` `lnf´ `` to be filled in; 1⇒`` `lnf´ `` and `` `g´ `` to be filled in
`` `b´ ``	full parameter row vector $\mathbf{b}=(\mathbf{b}_1,\mathbf{b}_2,\ldots,\mathbf{b}_E)$
`` `lnf´ ``	scalar to be filled in with overall $\ln L$
`` `g´ ``	row vector to be filled in with overall $\mathbf{g}=\partial \ln L/\partial \mathbf{b}$
`` `negH´ ``	argument to be ignored
`` `g1´ ``	variable optionally to be filled in with $\partial \ln \ell_j/\partial \mathbf{b}_1$
`` `g2´ ``	variable optionally to be filled in with $\partial \ln \ell_j/\partial \mathbf{b}_2$
...	

See Appendix C, **Method d1 checklist**.

Method d2 evaluators:

```
program define progname
        version 6
        args todo b lnf g negH [g1 [g2 ... ] ]
        tempvar theta1 theta2 ...
        mleval `theta1´ = `b´, eq(1)
        mleval `theta2´ = `b´, eq(2) /* if there is a θ2 */
        ...
        /* if you need to create any intermediate results: */
        tempvar tmp1 tmp2 ...
        gen double `tmp1´ = ...
        ...
        mlsum `lnf´ = ...
        if `todo´==0 | `lnf´==. { exit }
        tempname d1 d2 ...
        mlvecsum `lnf´ `d1´ = formula for ∂ ln ℓj/∂θ1j, eq(1)
        mlvecsum `lnf´ `d2´ = formula for ∂ ln ℓj/∂θ2j, eq(2)
        ...
        matrix `g´ = (`d1´, `d2´, ... )
        if `todo´==1 | `lnf´==. { exit }
        tempname d11 d12 d22 ...
        mlmatsum `lnf´ `d11´ = formula for −∂² ln ℓj/∂θ²1j, eq(1)
        mlmatsum `lnf´ `d12´ = formula for −∂² ln ℓj/∂θ1j∂θ2j, eq(1,2)
        mlmatsum `lnf´ `d22´ = formula for −∂² ln ℓj/∂θ²2j, eq(2)
        ...
        matrix `negH´ = (`d11´,`d12´,... \ `d12´´,`d22´,... )
end
```

where

`` `todo´ ``	contains 0, 1, or 2 0⇒`` `lnf´ `` to be filled in; 1⇒`` `lnf´ `` and `` `g´ `` to be filled in; 2⇒`` `lnf´ ``, `` `g´ ``, and `` `negH´ `` to be filled in
`` `b´ ``	full parameter row vector $\mathbf{b}=(\mathbf{b}_1,\mathbf{b}_2,\ldots,\mathbf{b}_E)$
`` `lnf´ ``	scalar to be filled in with overall $\ln L$

`g'	row vector to be filled in with overall $\mathbf{g}=\partial \ln L/\partial \mathbf{b}$
`negH'	matrix to be filled in with overall negative Hessian $-\mathbf{H}=-\partial^2 \ln L/\partial\mathbf{b}\partial\mathbf{b}'$
`g1'	variable optionally to be filled in with $\partial \ln \ell_j/\partial \mathbf{b}_1$
`g2'	variable optionally to be filled in with $\partial \ln \ell_j/\partial \mathbf{b}_2$
...	

See Appendix C, **Method d2 checklist**.

Global macros for use by all evaluators

$ML_y1	name of first dependent variable
$ML_y2	name of second dependent variable, if any
...	
$ML_samp	variable containing 1 if observation to be used; 0 otherwise
$ML_w	variable containing weight associated with observation or 1 if no weights specified

Method lf evaluators can ignore `$ML_samp`, but restricting calculations to the `$ML_samp==1` subsample will speed execution. Method lf evaluators must ignore `$ML_w`; application of weights is handled by the method itself.

Method d0, d1, and d2 can ignore `$ML_samp` as long as `ml model`'s `nopreserve` option is not specified. Method d0, d1, and d2 will run more quickly if `nopreserve` is specified. Method d0, d1, and d2 evaluators can ignore `$ML_w` only if they use `mlsum`, `mlvecsum`, and `mlmatsum` to produce final results.

Description

`ml clear` clears the current problem definition. This command is rarely if ever used because, when you type `ml model`, any previous problem is automatically cleared.

`ml model` defines the current problem. Equation syntax is described in **2.5 Specifying equations**. Maximization methods are described in **2.4 Choosing a maximization method**. Using `ml model` in a programming context is described in **13.5 ml model**.

`ml query` displays a description of the current problem. It is described in **9.2.3 Determining equation names**.

`ml check` verifies that the log-likelihood evaluator you have written seems to work. We strongly urge the use of this command. It is described in **8.2 ml check**.

`ml search` searches for (better) initial values. We recommend use of this command. It is described in **9.2 ml search**.

`ml plot` provides a graphical way of searching for (better) initial values. It is described in **9.3 ml plot**.

`ml init` provides of a way of setting initial values to user-specified values. It is described in **9.4 ml init**.

`ml report` reports the values of $\ln L$, its gradient, and its negative Hessian at the initial values or current parameter estimates $\mathbf{b}_0$. It is described in **10.3 Pressing the Break key**.

`ml trace` traces the execution of the user-defined log-likelihood evaluation program. It is described in **8.4 ml trace**.

`ml count` counts the number of times the user-defined log-likelihood evaluation program is called. It was intended as a debugging tool for those developing `ml` and now it serves little use besides entertainment. `ml clear` clears the counter. `ml count on` turns on the counter. `ml count` without arguments reports the current values of the counters. `ml count off` stops counting calls.

`ml maximize` maximizes the likelihood function and reports final results. It is described in **10 Interactive maximization**. Once `ml maximize` has successfully completed, the previously mentioned `ml` commands may no longer be used—`ml graph` and `ml display` may be used.

`ml graph` graphs the log-likelihood values against the iteration number. It is described in **11.2 Graphing convergence**.

`ml display` redisplays final results. It is described in **11.3 Redisplaying output**.

progname is the name of a program you write to evaluate the log-likelihood function. In this documentation, it is referred to as the user-written evaluator or sometimes simply as the evaluator. The program you write is written in the style required by the method you choose. The methods are lf, d0, d1, and d2. Thus, if you choose to use method lf, your program is called a method lf evaluator. Method lf evaluators are required to evaluate the observation-by-observation log likelihood $\ln\ell_j$, $j = 1, \ldots, N$. Method d0 evaluators are required to evaluate the overall log likelihood $\ln L$. Method d1 evaluators are required to evaluate the overall log likelihood and its gradient vector $\mathbf{g} = \partial \ln L/\partial \mathbf{b}$. Method d2 evaluators are required to evaluate the overall log likelihood, its gradient, and its negative Hessian matrix $-H = -\partial^2 \ln L/\partial\mathbf{b}\partial\mathbf{b}'$.

The program you are to write is described in **3 Method lf**, **5 Method d0**, **6 Method d1**, or **7 Method d2** depending on the method chosen.

`mleval` is a subroutine for use by method d0, d1, and d2 evaluators to evaluate the coefficient vector $\mathbf{b}$ that they are passed. It is described in **4.2.3 Using mleval**.

`mlsum` is a subroutine for use by method d0, d1, and d2 evaluators to define the value $\ln L$ that is to be returned. It is described in **4.2.6 Using mlsum**.

`mlvecsum` is a subroutine for use by method d1 and d2 evaluators to define the gradient vector $\mathbf{g}$ that is to be returned. It is suitable for use only when the likelihood function meets the linear-form restrictions. It is described in **4.2.8 Using mlvecsum**.

`mlmatsum` is a subroutine for use by method d2 evaluators to define the negative Hessian $-\mathbf{H}$ matrix that is to be returned. It is suitable for use only when the likelihood function meets the linear-form restrictions. It is described in **4.2.10 Using mlmatsum**.

Options for use with ml model in interactive or noninteractive mode

`robust` and `cluster(`*varname*`)` specify the robust estimate of variance, as does specifying `pweights`.

If you have written a method lf evaluator, `robust`, `cluster()`, and `pweights` will work. There is nothing to do except specify the options.

If you have written a method d0 evaluator, `robust`, `cluster()`, and `pweights` will not work. Specifying these options will result in an error message.

If you have written a method d1 or d2 evaluator and the likelihood function satisfies the linear-form restrictions, `robust`, `cluster()`, and `pweights` will work only if you fill in the equation scores—see **6.3 Robust variance estimates**—otherwise specifying these options will result in an error message.

`title(`*string*`)` specifies the title to be placed on the estimation output when results are complete.

`nopreserve` specifies that it is not necessary for `ml` to ensure that only the estimation subsample is in memory when the user-written likelihood evaluator is called. `nopreserve` is irrelevant when using method lf.

For the other methods, if `nopreserve` is not specified, then `ml` saves the dataset in a file (preserves the original dataset) and drops the irrelevant observations before calling the user-written evaluator.

This way, even if the evaluator does not restrict its attentions to the `$ML_samp==1` subsample, results will still be correct. Later, `ml` automatically restores the original data.

`ml` need not go through these machinations in the case of method lf because the user-written evaluator calculates observation-by-observation values and it is `ml` itself that sums the components.

`ml` goes through these machinations if and only if the estimation sample is a subsample of the data in memory. If the estimation sample includes every observation in memory, `ml` does not preserve the original data. Thus, programmers must not damage the original data unless they `preserve` the data themselves.

We recommend interactive users of `ml` not specify `nopreserve`; the speed gain is not worth the chances of incorrect results.

We recommend programmers do specify `nopreserve`, but only after verifying that their evaluator really does restrict its attentions solely to the `$ML_samp==1` subsample.

`collinear` specifies that `ml` is not to remove the collinear variables within equations. There is no reason one would want to leave collinear variables in place, but this option is of interest to programmers who, in their code, have already removed collinear variables and thus do not want `ml` to waste computer time checking again.

`missing` specifies that observations containing variables with missing values are not to be eliminated from the estimation sample. There are two reasons one might want to specify `missing`.

Programmers may wish to specify `missing` because, in other parts of their code, they have already eliminated observations with missing values and thus do not want `ml` to waste computer time looking again.

All users may wish to specify `missing` if their model explicitly deals with missing values. Stata's `heckman` command is a good example of this. In such cases, there will be observations where missing values are allowed and other observations where they are not—where their presence should cause the observation to be eliminated. If you specify `missing`, it is your responsibility to specify an `if` *exp* that eliminates the irrelevant observations.

`lf0(`$\#_k$ $\#_{ll}$`)` is typically used by programmers. It specifies the number of parameters and log-likelihood value of the "constant-only" model so that `ml` can report a likelihood-ratio test rather than a Wald test. These values were, perhaps, analytically determined, or they may have been determined by a previous estimation of the constant-only model on the estimation sample.

Also see the `continue` option directly below.

If you specify `lf0()`, it must be safe for you to specify the `missing` option, too, else how did you calculate the log likelihood for the constant-only model on the same sample? You must have identified the estimation sample, and done so correctly, so there is no reason for `ml` to waste time rechecking your results. Which is to say, do not specify `lf0()` unless you are certain your code identifies the estimation sample correctly.

`lf0()`, even if specified, is ignored if `robust`, `cluster()`, or `pweights` are specified because in that case a likelihood-ratio test would be inappropriate.

`continue` is typically specified by programmers. It does two things:

First, it specifies that a model has just been estimated, by either `ml` or some other estimation command such as `logit`, and that the likelihood value stored in `e(ll)` and the number of parameters stored in `e(b)` as of this instant are the relevant values of the constant-only model. The current value of the log likelihood is used to present a likelihood-ratio test unless `robust`, `cluster()`, or `pweights` are specified because, in that case, a likelihood-ratio test would be inappropriate.

Second, `continue` sets the initial values $\mathbf{b}_0$ for the model about to be estimated according to the `e(b)` currently stored.

The comments made about specifying `missing` with `lf0()` apply equally well in this case.

`waldtest(#)` is typically specified by programmers. By default, `ml` presents a Wald test, but that is overridden if options `lf0()` or `continue` are specified, and that is overridden again (so we are back to the Wald test) if `robust`, `cluster()`, or `pweights` are specified.

`waldtest(0)` prevents even the Wald test from being reported.

`waldtest(-1)` is the default. It specifies that a Wald test is to be performed, if it is performed, by constraining all coefficients except for the intercept to 0 in the first equation. Remaining equations are to be unconstrained. The logic as to whether a Wald test is performed is the standard: perform the Wald test if neither `lf0()` nor `continue` were specified, but force a Wald test if `robust`, `cluster`, or `pweights` were specified.

`waldtest(`*k*`)` for $k \leq -1$ specifies that a Wald test is to be performed, if it is performed, by constraining all coefficients except for intercepts to 0 in the first $|k|$ equations; remaining equations are to be unconstrained. The logic as to whether a Wald test is performed is the standard.

`waldtest(`*k*`)` for $k \geq 1$ works like the above except that it forces a Wald test to be reported even if the information to perform the likelihood-ratio test is available and even if none of `robust`, `cluster`, or `pweights` were specified. `waldtest(`*k*`)`, $k \geq 1$, may not be specified with `lf0()`.

`obs(#)` is used mostly by programmers. It specifies that the number of observations reported, and ultimately stored in `e(N)`, is to be #. Ordinarily, `ml` works that out for itself, and correctly. Programmers may want to specify this option when, in order for the likelihood-evaluator to work for N observations, they first had to modify the data so that it contained a different number of observations.

`noscvars` is used mostly by programmers. It specifies that method d0, d1, or d2 is being used but that the likelihood evaluation program does not calculate nor use arguments `` `g1' ``, `` `g2' ``, etc., which are the score vectors; see **6.3 Robust variance estimates**. Thus, `ml` can save a little time by not generating and passing those arguments.

Options for use with ml model in noninteractive mode

In addition to the above options, the following options are for use with `ml model` in noninteractive mode. Noninteractive mode is for programmers who use `ml` as a subroutine and want to issue a single command that will carry forth the estimation from start to finish; see **13.5 ml model**.

`maximize` is not optional. It specifies noninteractive mode.

`init(`*ml_init_args*`)` sets the initial values $\mathbf{b}_0$. *ml_init_args* are whatever you would type after the `ml init` command.

`search(on|quietly|off)` specifies whether `ml search` is to be used to improve the initial values. `search(on)` is the default and is equivalent to running separately "`ml search, repeat(0)`". `search(quietly)` is the same as `search(on)` except that it suppresses `ml search`'s output. `search(off)` prevents the calling of `ml search` altogether.

`repeat(#)` is `ml search`'s `repeat()` option and is relevant only if `search(off)` is not specified. `repeat(0)` is the default.

`bounds(`*ml_search_bounds*`)` is relevant only if `search(off)` is not specified. The command `ml model` issues is "`ml search` *ml_search_bounds*, `repeat(#)`". Specifying search bounds is optional.

`difficult`, `nolog`, `trace`, `gradient`, `hessian`, `showstep`, `iterate()`, `ltolerance()`, `tolerance()`, `nowarning`, `novce`, and `score()` are `ml maximize`'s equivalent options.

Options for use when specifying equations

`noconstant` specifies that the equation is not to include an intercept.

`offset(`*varname*`)` specifies that the equation is to be **xb** + *varname*; that the equation is to include *varname* with coefficient constrained to be 1.

`exposure(`*varname*`)` is an alternative to `offset(`*varname*`)`; it specifies that the equation is to be **xb** + ln(*varname*). The equation is to include ln(*varname*) with coefficient constrained to be 1.

Options for use with ml search

`repeat(`*#*`)` specifies the number of random attempts that are to be made to find a better initial-value vector. The default is `repeat(10)`.

`repeat(0)` specifies that no random attempts are to be made. More correctly, `repeat(0)` specifies that no random attempts are to be made if the initial initial-value vector is a feasible starting point. If it is not, `ml search` will make random attempts even if you specify `repeat(0)` because it has no alternative. The `repeat()` option refers to the number of random attempts to be made to improve the initial values. When the initial starting value vector is not feasible, `ml search` will make up to 1,000 random attempts to find starting values. It stops the instant it finds one set of values that works and then moves into its improve-initial-values logic.

`repeat(`k`)`, $k > 0$, specifies the number of random attempts to be made to improve the initial values.

`nolog` specifies that no output is to appear while `ml search` looks for better starting values. If you specify `nolog` and the initial starting-value vector is not feasible, `ml search` will ignore the fact that you specified the `nolog` option. If `ml search` must take drastic action to find starting values, it feels you should know about this even if you attempted to suppress its usual output.

`trace` specifies that you want more detailed output about `ml search`'s actions than it would usually provide. This is more entertaining than useful. `ml search` prints a period each time it evaluates the likelihood function without obtaining a better result and a plus when it does.

`restart` specifies that random actions are to be taken to obtain starting values and that the resulting starting values are not to be a deterministic function of the current values. Users should not specify this option mainly because, with `restart`, `ml search` intentionally does not produce as good a set of starting values as it could. `restart` is included for use by the optimizer when it gets into serious trouble. The random actions are to ensure that the actions of the optimizer and `ml search`, working together, do not result in a long, endless loop.

`restart` implies `norescale`, which is why we recommend you do not specify `restart`. In testing, cases were discovered where `rescale` worked so well that, even after randomization, the rescaler would bring the starting values right back to where they had been the first time and so defeated the intended randomization.

`norescale` specifies that `ml search` is not to engage in its rescaling actions to improve the parameter vector. We do not recommend specifying this option because rescaling tends to work so well.

Options for use with ml plot

`saving(`*filename*[`, replace`]`)` specifies that the graph is to be saved in *filename*`.gph`.

Options for use with ml init

`skip` specifies that any parameters found in the specified initialization vector that are not also found in the model are to be ignored. The default action is to issue an error message.

`copy` specifies that the list of numbers or the initialization vector is to be copied into the initial-value vector by position rather than by name.

Options for use with ml maximize

`difficult` specifies that the likelihood function is likely to be difficult to maximize. In particular, `difficult` states that there may be regions where $-\mathbf{H}$ is not invertible and that, in those regions, `ml`'s standard fixup may not work well. `difficult` specifies that a different fixup requiring substantially more computer time is to be used. For the majority of likelihood functions, `difficult` is likely to increase execution times unnecessarily. For other likelihood functions, specifying `difficult` is of great importance; see **10.4 Maximizing difficult likelihood functions**.

`nolog`, `trace`, `gradient`, `hessian`, and `showstep` control the display of the iteration log.

`nolog` suppresses reporting of the iteration log.

`trace` adds to the iteration log a report on the current parameter vector.

`gradient` adds to the iteration log a report on the current gradient vector.

`hessian` adds to the iteration log a report on the current negative Hessian matrix.

`showstep` adds to the iteration log a report on the steps within iteration.

`iterate(`*#*`)`, `ltolerance(`*#*`)`, and `tolerance(`*#*`)` specify the definition of convergence.

`iterate(16000) tolerance(1e-6) ltolerance(1e-7)` are the default.

Convergence is declared when

$$\texttt{mreldif}(\mathbf{b}_{i+1}, \mathbf{b}_i) \le \texttt{tolerance()}$$

$$\textbf{or} \quad \texttt{reldif}(\ln L(\mathbf{b}_{i+1}), \ln L(\mathbf{b}_i)) \le \texttt{ltolerance()}$$

In addition, iteration stops when $i =$ `iterate()`; in that case, results along with the message "convergence not achieved" are presented. The return code is still set to 0.

`nowarning` is allowed only with `iterate(0)`. `nowarning` suppresses the "convergence not achieved" message. Programmers might specify `iterate(0) nowarning` when they have a vector $\mathbf{b}$ already containing what are the final estimates and want `ml` to calculate the variance matrix and post final estimation results. In that case, specify "`init(b) search(off) iterate(0) nowarning nolog`".

`novce` is allowed only with `iterate(0)`. `novce` substitutes the zero matrix for the variance matrix which in effect posts estimation results as fixed constants.

`score(`*newvarlist*`)` specifies that the equation scores are to be stored in the specified new variables. Either specify one new variable name per equation or specify a short name suffixed with a `*`. E.g., `score(sc*)` would be taken as specifying `sc1` if there were one equation and `sc1` and `sc2` if there were two equations. In order to specify `score()`, either you must be using method lf or the estimation subsample must be the entire data in memory or you must have specified the `nopreserve` option.

`nooutput` suppresses display of the final results. This is different from prefixing `ml maximize` with `quietly` in that the iteration log is still displayed (assuming `nolog` is not specified).

`level(`*#*`)` is the standard confidence-level option. It specifies the confidence level, in percent, for confidence intervals of the coefficients.

`eform(`*string*`)` is `ml display`'s `eform()` option.

`noclear` specifies that after the model has converged, the ml problem definition is not to be cleared. Perhaps you are having convergence problems and intend to run the model to convergence, then use `ml search` to see if those values can be improved, and then start the estimation again.

Options for use with ml graph

`saving(`*filename*[`, replace`]`)` specifies that the graph is to be saved in *filename*`.gph`.

Options for use with ml display

`noheader` suppresses display of the header above the coefficient table that displays the final log-likelihood value, the number of observations, and the model significance test.

`eform(`*string*`)` displays the coefficient table in exponentiated form: for each coefficient, $\exp(b)$ rather than b is displayed and standard errors and confidence intervals are transformed. Display of the intercept, if any, is suppressed. *string* is the table header that will be displayed above the transformed coefficients and must be 11 characters or less in length, for example, `eform("Odds ratio")`

`first` displays a coefficient table reporting results for the first equation only, and the report makes it appear that the first equation is the only equation. This is used by programmers who estimate ancillary parameters in the second and subsequent equations and will report the values of such parameters themselves.

`neq(`*#*`)` is an alternative to `first`. `neq(`*#*`)` displays a coefficient table reporting results for the first *#* equations. This is used by programmers who estimate ancillary parameters in the $\# + 1$ and subsequent equations and will report the values of such parameters themselves.

`plus` displays the coefficient table just as it would be ordinarily, but then, rather than ending the table in a line of dashes, ends it in dashes–plus-sign–dashes. This is so that programmers can write additional display code to add more results to the table and make it appear as if the combined result is one table. Programmers typically specify `plus` with options `first` or `neq()`.

`level(`*#*`)` is the standard confidence-level option. It specifies the confidence level, in percent, for confidence intervals of the coefficients.

Options for use with mleval

`eq(`*#*`)` specifies the equation number i for which $\theta_{ij} = \mathbf{x}_{ij}\mathbf{b}_i$ is to be evaluated. `eq(1)` is assumed if `eq()` is not specified.

`scalar` asserts that the ith equation is known to evaluate to a constant; the equation was specified as `()`, `(`*name*`:)`, or `/`*name* on the `ml model` statement. If you specify this option, the new variable created is created as a scalar. If the ith equation does not evaluate to a scalar, an error message is issued.

Options for use with mlsum

`noweight` specifies that weights (`$ML_w`) are to be ignored when summing the likelihood function.

Options for use with mlvecsum

`eq(`*#*`)` specifies the equation for which a gradient vector $\partial \ln L/\partial \mathbf{b}_i$ is to be constructed. The default is `eq(1)`.

Options for use with mlmatsum

`eq(`*#*[`,`*#*]`)` specifies the equations for which the negative Hessian matrix is to be constructed. The default is `eq(1)`, which means the same as `eq(1,1)`, which means $-\partial^2 \ln L/\partial \mathbf{b}_1 \partial \mathbf{b}_1'$. Specifying `eq(`*i*`,`*j*`)` results in $-\partial^2 \ln L/\partial \mathbf{b}_i \partial \mathbf{b}_j'$.

B Method lf checklist

1. In order to use method lf, the likelihood function must meet the linear-form restrictions; otherwise, you must use method d0.

2. Write a program to evaluate $\ln \ell_j$, the contribution to the overall likelihood for an observation. The outline for the program is

```
program define myprog
        version 6
        args lnf theta1 theta2 ...
        /* if you need to create any intermediate results: */
        tempvar tmp1 tmp2 ...
        quietly gen double `tmp1´ = ...
        ...
        quietly replace `lnf´ = ...
end
```

   Access the dependent variables (if any) by typing `$ML_y1`, `$ML_y2`, ... Use `$ML_y1`, `$ML_y2`, ..., just as you would any existing variable in the data.

   Final results are to be saved in `` `lnf´ ``. (`` `lnf´ `` is just a double-precision variable `ml` created for you. When `ml` calls your program, `` `lnf´ `` contains missing values.)

   It is of great importance that, if you create any temporary variables for intermediate results, you create them as `doubles`: Type "`gen double` *name* `=` ..."; do not omit the word `double`.

3. Issue the `ml model` statement as

```
. ml model lf myprog (equation for θ1) (equation for θ2) ...
```

   Specify any `if` *exp*, weights, or the `robust` and/or `cluster()` option at this point.

4. Verify your program works; type

```
. ml check
```

5. Find starting values; type

```
. ml search
```

   or use `ml init` or `ml plot`. We suggest you use `ml search`. In most cases, the `ml search` can be omitted, too, because `ml maximize` calls `ml search`, but in that case `ml search` will not do as thorough a job.

6. Obtain estimates; type

```
. ml maximize
```

7. If later, you wish to redisplay results, type

```
. ml display
```

C Method d0 checklist

1. Method d0 may be used with any likelihood function.
2. Write a program to evaluate $\ln L$, the overall log-likelihood function. The outline for the evaluator is

   ```
   program define myprog
           version 6
           args todo b lnf

           tempvar theta1 theta2 ...
           mleval `theta1´ = `b´
           mleval `theta2´ = `b´, eq(2) /* if there is a θ2 */
           ...

           /* if you need to create any intermediate results: */
           tempvar tmp1 tmp2 ...
           gen double `tmp1´ = ...
           ...

           mlsum `lnf´ = ...
   end
   ```

   Argument `` `todo´ `` will always contain `0`; it can be ignored.

   You are passed the parameter row vector in `` `b´ ``. Use `mleval` to obtain the thetas from it.

   Access the dependent variables (if any) by typing `$ML_y1`, `$ML_y2`, ... Use `$ML_y1`, `$ML_y2`, ..., just as you would any existing variable in the data.

   Final results are to be saved in scalar `` `lnf´ ``. Use `mlsum` to produce it.

   It is of great importance that, if you create any temporary variables for intermediate results, you create them as `doubles`: Type "`gen double` *name* `= ...`"; do not omit the word `double`.

   The estimation subsample is `$ML_samp==1`. You may safely ignore this if you do not specify `ml model`'s `nopreserve` option. When your program is called, only relevant data will be in memory. If you do specify `nopreserve`, understand that `mleval` and `mlsum` automatically restrict themselves to the estimation subsample; it is not necessary to code `if $ML_samp==1` on these commands. You merely need to restrict other commands to the `$ML_samp==1` subsample.

   The weights are stored in variable `$ML_w`, which contains 1 in every observation if no weights are specified. If you use `mlsum` to produce the log likelihood and `mlvecsum` to produce the gradient vector, you may ignore this because these commands handle that themselves.
3. Issue the `ml model` statement as

   ```
   . ml model d0 myprog (equation for θ1) (equation for θ2) ...
   ```

   Specify any `if` *exp* or weights at this point. While method d0 evaluators can be used with `fweights`, `aweights`, and `iweights`, `pweights` may not be specified. Use method lf or method d1 if this is important. Similarly, you may not specify `ml model`'s `robust` or `cluster()` options. Use method lf or method d1 if this is important.
4. Verify your program works; type

   ```
   . ml check
   ```

5. Find starting values; type

```
. ml search
```

or use `ml init` or `ml plot`. We suggest you use `ml search`. In most cases, the `ml search` can be omitted, too, because `ml maximize` calls `ml search`, but in that case `ml search` will not do as thorough a job.

6. Obtain estimates; type

```
. ml maximize
```

7. If later, you wish to redisplay results, type

```
. ml display
```

D Method d1 checklist

1. Method d1 may be used with any likelihood function. This will be a considerable effort, however, if the function does not meet the linear-form restrictions.

2. Write a program to evaluate $\ln L$, the overall log-likelihood function, and its derivatives. The outline for the evaluator is

```
program define myprog
        version 6
        args todo b lnf g
        tempvar theta1 theta2 ...
        mleval `theta1' = `b'
        mleval `theta2' = `b', eq(2) /* if there is a θ2 */
        ...
        /* if you need to create any intermediate results: */
        tempvar tmp1 tmp2 ...
        gen double `tmp1' = ...
        ...
        mlsum `lnf' = ...
        if `todo'==0 | `lnf'==. { exit }
        tempname d1 d2 ...
        mlvecsum `lnf' `d1' = formula for ∂ln ℓj/∂θ1j, eq(1)
        mlvecsum `lnf' `d2' = formula for ∂ln ℓj/∂θ2j, eq(2)
        ...
        matrix `g' = (`d1', `d2', ... )
end
```

Argument `` `todo' `` will contain `0` or `1`. You are to fill in the log likelihood if `` `todo'==0 `` or the log likelihood and gradient otherwise.

You are passed the parameter row vector in `` `b' ``. Use **mleval** to obtain the thetas from it.

Access the dependent variables (if any) by referring to **$ML_y1**, **$ML_y2**, ... Use **$ML_y1**, **$ML_y2**, ..., just as you would any existing variable in the data.

The overall log-likelihood value is to be saved in scalar `` `lnf' ``. Use `mlsum` to produce it.

The gradient vector is to be saved in vector `` `g' ``. Use `mlvecsum` to produce it. You issue one `mlvecsum` per theta and then use `matrix` *name* `= ...` to put the results together.

It is of great importance that, if you create any temporary variables for intermediate results, you create them as `doubles`: Type "`gen double` *name* `= ...`"; do not omit the word `double`.

The estimation subsample is `$ML_samp==1`. You may safely ignore this if you do not specify `ml model`'s `nopreserve` option. When your program is called, only relevant data will be in memory. If you do specify `nopreserve`, understand that `mleval`, `mlsum`, and `mlvecsum` automatically restrict themselves to the estimation subsample; it is not necessary to code `if $ML_samp==1` on these commands. You merely need to restrict other commands to the `$ML_samp==1` subsample.

The weights are stored in variable `$ML_w`, which contains 1 if no weights are specified. If you use `mlsum` to produce the log likelihood and `mlvecsum` to produce the gradient vector, you may ignore this because these commands handle the weights themselves.

3. Issue the `ml model` statement as

   . `ml model d1debug` *myprog* (*equation for* θ_1) (*equation for* θ_2) ...

   Note well: begin with method d1debug and not method d1. This tests whether the program produces correct first derivatives.

4. Verify the program works, type

   ```
   . ml check
   ```

5. Find starting values; type

   ```
   . ml search
   ```

   or use `ml init` or `ml plot`. We suggest you use `ml search`. In most cases, the `ml search` can be omitted, too, because `ml maximize` calls `ml search`, but in that case `ml search` will not do as thorough a job.

6. Obtain estimates; type

   ```
   . ml maximize
   ```

   Review the reported `mreldif()`s in the iteration log. Even if you have programmed the derivatives correctly, it is typical to observe the `mreldif()`s falling and then rising and even observing the `mreldif()` being largest in the final iteration, when the gradient vector is near zero. This should not concern you. The reported `mreldif()`s should be 10^{-3} or smaller in the middle iterations. If you suspect problems, reissue the `ml model` statement and then run `ml maximize` with the `gradient` option.

7. Now that you have verified that your program is calculating derivatives correctly, switch to using method d1 by typing

   . `ml model d1` *myprog* (*equation for* θ_1) (*equation for* θ_2) ...

   Repeat Steps 4, 5, and 6. You should obtain approximately the same results. Note that the iteration log will probably differ; method d1 may require fewer or more iterations than method d1debug. This does not indicate problems. Method d1debug used the numerically calculated derivatives and maximization is a chaotic process.

8. If you wish to use `pweights` or `ml model`'s `robust` or `cluster()` options, change the `args` statement at the top of the program from

   ```
   args todo b lnf g
   ```

   to

   ```
   args todo b lnf g negH g1 g2 ...
   ```

   Change the part of the outline that reads

   ```
   tempname d1 d2 ...
   mlvecsum `lnf' `d1' = formula for ∂ln ℓj/∂θ1j, eq(1)
   mlvecsum `lnf' `d2' = formula for ∂ln ℓj/∂θ2j, eq(2)
   ...
   ```

   to read

   ```
   quietly replace `g1' = formula for ∂ln ℓj/∂θ1j
   quietly replace `g2' = formula for ∂ln ℓj/∂θ2j
   ...
   tempname d1 d2 ...
   mlvecsum `lnf' `d1' = `g1', eq(1)
   mlvecsum `lnf' `d2' = `g2', eq(2)
   ...
   ```

You can do this only if your likelihood function meets the linear-form restrictions.

9. If later, you wish to redisplay results, type

```
. ml display
```

E Method d2 checklist

1. Method d2 may be used with any likelihood function. This will be a considerable effort, however, if the function does not meet the linear-form restrictions.

2. Write a program to evaluate $\ln L$, the overall log-likelihood function, and its first and second derivatives. The outline for the evaluator is

```
program define myprog
        version 6
        args todo b lnf g negH
        tempvar theta1 theta2 ...
        mleval `theta1´ = `b´
        mleval `theta2´ = `b´, eq(2) /* if there is a θ2 */
        ...
        /* if you need to create any intermediate results: */
        tempvar tmp1 tmp2 ...
        gen double `tmp1´ = ...
        ...
        mlsum `lnf´ = ...
        if `todo´==0 | `lnf´==. { exit }
        tempname d1 d2 ...
        mlvecsum `lnf´ `d1´ = formula for ∂ ln ℓj/∂θ1j, eq(1)
        mlvecsum `lnf´ `d2´ = formula for ∂ ln ℓj/∂θ2j, eq(2)
        ...
        matrix `g´ = (`d1´, `d2´, ... )
        if `todo´==1 | `lnf´==. { exit }
        tempname d11 d12 d22 ...
        mlmatsum `lnf´ `d11´ = formula for −∂2 ln ℓj/∂θ1j^2, eq(1)
        mlmatsum `lnf´ `d12´ = formula for −∂2 ln ℓj/∂θ1j∂θ2j, eq(1,2)
        mlmatsum `lnf´ `d22´ = formula for −∂2 ln ℓj/∂θ2j^2, eq(2)
        ...
        matrix `negH´ = (`d11´,`d12´,... \ `d12´´,`d22´,... )
end
```

Argument `` `todo´ `` will contain `0`, `1`, or `2`. You are to fill in the log likelihood if `` `todo´==0 ``; the log likelihood and gradient if `` `todo´==1 ``; or the log likelihood, gradient, and Hessian otherwise.

You are passed the parameter row vector in `` `b´ ``. Use **mleval** to obtain the thetas from it.

Access the dependent variables (if any) by referring to `$ML_y1`, `$ML_y2`, ... Use `$ML_y1`, `$ML_y2`, ..., just as you would any existing variable in the data.

The overall log-likelihood value is to be saved in scalar `` `lnf´ ``. Use **mlsum** to produce it.

The gradient vector is to be saved in vector `` `g´ ``. Use **mlvecsum** to produce it. You issue one **mlvecsum** per theta and then use **matrix** *name* = ... to put the results together.

The negative Hessian is to be saved in matrix `` `negH´ ``. Use **mlmatsum** to produce it. You issue one **mlmatsum** per equation and equation pair and then use **matrix** *name* = ... to put the results together.

It is of great importance that, if you create any temporary variables for intermediate results, you create them as `double`s: Type "`gen double` *name* = ..."; do not omit the word `double`.

The estimation subsample is `$ML_samp==1`. You may safely ignore this if you do not specify **ml model**'s **nopreserve** option. When your program is called, only relevant data will be in memory. If you do specify `nopreserve`, understand that `mleval`, `mlsum`, `mlvecsum`, and `mlmatsum` automatically restrict themselves to the estimation subsample; it is not necessary to code `if $ML_samp==1` on these commands. You merely need to restrict other commands to the `$ML_samp==1` subsample.

The weights are stored in variable `$ML_w`, which contains 1 if no weights are specified. If you use `mlsum` to produce the log likelihood and `mlvecsum` to produce the gradient vector, you may ignore this because these commands handle the weights themselves.

3. Issue the `ml model` statement as

   . ml model d2debug *myprog* (*equation for* θ_1) (*equation for* θ_2) ...

   Note well, begin with method d2debug and not method d2. This allows testing whether the program produces correct first and second derivatives.

4. Verify the program works; type

```
. ml check
```

5. Find starting values; type

```
. ml search
```

   or use `ml init` or `ml plot`. We suggest you use `ml search`. In most cases, the `ml search` can be omitted, too, because `ml maximize` calls `ml search`, but in that case `ml search` will not do as thorough a job.

6. Obtain estimates; type

```
. ml maximize
```

   Review the reported `mreldif()`s in the iteration log. Even if you have programmed the derivatives correctly, it is typical to observe the `mreldif()`s for the gradient vector falling and then rising and even observing the `mreldif()` being largest in the final iteration, when the gradient vector is near zero. This should not concern you. The reported `mreldif()`s should be 1e–3 or smaller in the middle iterations.

   For the negative Hessian, early iterations often report large `mreldif()`s but the `mreldifs()` fall. The reported `mlreldif` should be 1e–5 or smaller in the last iteration.

   If you suspect problems, reissue the `ml model` statement and then run `ml maximize` with the `gradient` option and/or `hessian` option.

7. Now that you have verified that your program is calculating derivatives correctly, switch to using method d2 by typing

   . ml model d2 *myprog* (*equation for* θ_1) (*equation for* θ_2) ...

   Repeat steps 4, 5, and 6. You should obtain approximately the same results (relative differences being 1e–8 for coefficients and smaller than 1e–5 for standard errors. Note that the iteration log will probably differ; method d2 may require fewer or more iterations than method d2debug. This does not indicate problems. Method d2debug used the numerically calculated derivatives and maximization is a chaotic process.

9. If you wish to use `pweights` or `ml model`'s `robust` or `cluster()` options, change the `args` statement at the top of the program from

```
args todo b lnf g negH
```

   to

```
args todo b lnf g negH g1 g2 ...
```

Change the part of the outline that reads

```
tempname d1 d2 ...
mlvecsum `lnf´ `d1´ = formula for ∂ ln ℓj/∂θ1j, eq(1)
mlvecsum `lnf´ `d2´ = formula for ∂ ln ℓj/∂θ2j, eq(2)
...
```

to read

```
quietly replace `g1´ = formula for ∂ ln ℓj/∂θ1j
quietly replace `g2´ = formula for ∂ ln ℓj/∂θ2j
...
tempname d1 d2 ...
mlvecsum `lnf´ `d1´ = `g1´, eq(1)
mlvecsum `lnf´ `d2´ = `g2´, eq(2)
...
```

You can do this only if your likelihood function meets the linear-form restrictions.

9. If later, you wish to redisplay results, type

```
. ml display
```

Index